volume 46 2009

Essays in Biochemistry

The Polyamines: Small Molecules in the 'Omics' Era

Edited by H. Wallace

Portland Press

Essays in Biochemistry is published by Portland Press Ltd on behalf of the Biochemical Society

Portland Press Limited
Third Floor, Eagle House
16 Procter Street
London WC1V 6NX
U.K.
Tel.: +44 (0)20 7280 4110
Fax: +44 (0)20 7280 4169
email: editorial@portlandpress.com
www.portlandpress.com

British Library Cataloguing-in-Publication Data
A catalogue record for this book is available from the British Library
ISBN 978-1-85578-175-7
ISSN 0071 1365

Typeset by Aptara Inc., New Delhi, India
Printed in Great Britain by Latimer Trend Ltd, Plymouth

Essays in Biochemistry

Other recent titles in the Essays in Biochemistry series

Systems Biology: volume 45
edited by O. Wolkenhauer, P. Wellstead and K.-H. Cho
2008
ISBN 978 1 85578 170 2

Drugs and Ergogenic Aids to Improve Sport Performance: volume 44
edited by C.E. Cooper and R. Beneke
2008
ISBN 978 1 85578 1658

Oxygen Sensing and Hypoxia-Induced Responses: volume 43
edited by C. Peers
2007
ISBN 978 1 85578 160 3

The Biochemical Basis of the Health Effects of Exercise: volume 42
edited by A.J.M. Wagenmakers
2006
ISBN 978 1 85578 159 7

The Ubiquitin–Proteasome System: volume 41
edited by R.J. Mayer and R. Layfield
2005
ISBN 978 1 85578 153 5

The Nuclear Receptor Superfamily: volume 40
edited by I.J. McEwan
2004
ISBN 978 1 85578 150 4

Programmed Cell Death: volume 39
edited by T.G. Cotter
2003
ISBN 978 1 85578 148 1

Proteases in Biology and Medicine: volume 38
edited by N.M. Hooper
2002
ISBN 978 1 85578 147 4

Contents

Preface

The aim of this volume is to provide background information on the role of the polyamines in mammalian cells as we understand it currently in a concise and easily read format. Over the years, I have had many PhD, MSc, and intercalating medical and honours students who would have benefited enormously from a volume such as this to give them the initial understanding of the field that they need to start their research projects. This volume is dedicated to all of the students who have worked in my laboratory; I am sorry this took me so long to do! My thanks go to every one of you for all your hard work, dedication and fun over the years.

Why the title? We are now in an age where unparalleled amounts of data are being generated in biological systems – so much information that we have to design bioinformatics programs to deal with the data generated. This *Essays in Biochemistry* volume aims to show that there is still a place in the 'omics' world for elegant biochemistry which can lead to a detailed understanding of a biological system.

Heather Wallace
August 2009

Authors

Heather M. Wallace graduated in Biochemistry from the University of Glasgow and obtained her Ph.D. from the University of Aberdeen, Scotland, U.K. She undertook postdoctoral research in Biochemistry at Aberdeen and was awarded a University Research Fellowship. Her first Faculty position was a Wellcome lectureship followed by a "New Blood" lectureship/senior lectureship in Molecular Pharmacology and Toxicology. In 2004, she was elected a Fellow of the Royal College of Pathologists and currently is Chair of the College Specialty Advisory Committee for Toxicology. Heather was awarded a Fellowship of the British Toxicological Society in 2005 and became a member of the U.K. Register of Toxicologists and a European Registered Toxicologist in 2006. She is on the Editorial Board and is a Deputy Chair of the *Biochemical Journal* and reviews for many journals and grant awarding agencies. She is a registered evaluator for the European Commission. Her research interests are in the mechanisms of cell death associated with anticancer drugs and potential chemopreventative agents, in particular drugs targeting the polyamine pathway, and in the use of biomarkers for diagnosis and monitoring efficacy of anticancer drug therapy.

Lo Persson obtained a Ph.D. in Physiology from Lund University, Lund, Sweden in 1982. During 1983, he worked as a postdoctoral fellow at the Department of Cellular and Molecular Physiology, Milton S. Hershey Medical Center, Pennsylvania State University, PA, U.S.A. He is currently Professor of Molecular and Cellular Physiology at the Department of Experimental Medical Science, Lund University. He has worked in the field of polyamine-related research for more than 30 years. His research interests include the regulatory mechanisms involved in the cellular control of polyamine levels, as well as the polyamine metabolic pathway as a potential target for drug development against parasitic diseases.

Anthony Pegg is the J. Lloyd Huck Professor of Molecular and Cell Biology and Evan Pugh Professor of Cellular and Molecular Physiology at Pennsylvania State University College of Medicine in Hershey, PA, U.S.A. He received his Ph.D. from Cambridge University, Cambridge, U.K. in 1966 and was a postdoctoral fellow at Johns Hopkins University, Baltimore, MD, U.S.A., from 1966–68 where he worked with Guy Williams-Ashman and characterized the enzymes in the mammalian biosynthetic pathway for polyamines. His laboratory works on polyamine metabolism and function, and on DNA repair of alkylation damage. He has published more than 500 papers on these topics.

Chaim Kahana obtained a B.Sc. in Agriculture at the Hebrew University, Jerusalem, Israel, in 1976 and a M.Sc. in Agriculture in 1978. He studied towards his Ph.D. at the Weizmann Institute of Science, Rehovot, Israel, followed by postdoctoral research at the Johns Hopkins University (Baltimore, MD, U.S.A.) with Daniel Nathans. He is currently an Associate Professor in the Department of Molecular Genetics at the Weizmann Institute of Science, with interests in molecular biology, protein degradation, regulation of polyamine metabolism and in the role polyamines exert in regulating cellular functions, particularly cell growth and proliferation.

Kersti Alm has been involved in research on polyamines and their effects on the cell cycle since she was a graduate student. She has a temporary position at the level of Associate Professor at Lund University, Lund, Sweden, which includes teaching and research. Dr Alm has been a postdoctoral fellow at Roswell Park Cancer Institute, NY, U.S.A. Here she delved deeper into the mysteries of cell-cycle regulation together with Professor Carl Porter and Dr Deborah Kramer. **Stina M. Oredsson** has been in the field of cancer cells and cell-cycle regulation since 1979. She has always been fascinated by the possibilities inherent in cellular polyamine homoeostasis for utilization in cancer treatment. In later years her research has branched into studies of treating breast cancer, bladder cancer and neuroblastoma with polyamine analogues. Besides research, she is heavily involved in teaching at Lund University.

Thulani Senanayake is currently a Postdoctoral Associate at the University of Nebraska Medical Center in Omaha, NE, U.S.A. She was instrumental in developing an efficient synthesis of polyamine-based histone deacetylase inhibitors, and also conducted their evaluation as inhibitors of individual histone deacetylases, and as antitumour agents in cell culture. Her work led to the discovery of a relationship between histone deacetylase inhibition and annexin A1-mediated apoptosis in breast tumour cells. **Hemali Anunugama** is a Ph.D. graduate from the laboratory of Patrick Woster. As a graduate student in the Woster laboratory, Hemali was responsible for the synthesis of a series of polyamine–metal complexes, and for characterizing their DNA-binding and antitumour effects. **Tracey D. Boncher** is Associate Professor of Medicinal Chemistry in the College of Pharmacy at Ferris State University, MI, U.S.A. As a graduate student in the laboratory of Patrick Woster, Tracey synthesized an extensive library of symmetrically and unsymmetrically substituted polyamine analogues, many of which have been shown to possess significant antitumour activity. **Patrick M. Woster** is Professor of Pharmaceutical Sciences in the Eugene Applebaum College of Pharmacy and Health Sciences at Wayne State University. Professor Woster is a medicinal chemist with an interest in the synthesis of molecules that modulate polyamine metabolism as potential antitumour agents. He has produced a number of inhibitors that target enzymes in the polyamine biosynthetic pathway, and synthesized the first unsymmetrically substituted alkylpolyamine analogues. Molecules developed in the Woster laboratory have been shown to produce dramatic effects on a variety of tumour cells by initiating apoptosis, binding to DNA and by producing epigenetic changes in gene expression. Professor Woster currently serves as a consultant to Progen Pharmaceuticals.

Yi Huang is a Research Associate of Oncology in the Johns Hopkins University School of Medicine, Baltimore, MD, U.S.A. After completing his Ph.D. degree at the Medical University of South Carolina in 2001, Dr Huang joined the Breast Cancer Programme at Johns Hopkins for his postdoctoral training to study the role of polyamines in breast cancer development and the exploitation of polyamine analogues as an effective strategy in breast cancer therapy. In 2006, Dr Huang joined the faculty of the Johns Hopkins University School of Medicine to study the epigenetic regulation of histone-modifying enzymes on gene regulation and to develop effective polyamine analogues in targeting epigenetic alterations. **Laurence J. Marton** is the Chief Scientific Officer of Progen Pharmaceuticals and an Adjunct Professor in the Department of Laboratory Medicine at UCSF (University of California at San Francisco), School of Medicine During many years in academia, Professor Marton has served as Professor and Chair of the Department of Laboratory Medicine and as Professor of Neurological Surgery at UCSF, and subsequently as Dean of the Medical School at the University of Wisconsin, Madison, WI, U.S.A. More recently he co-founded a biotechnology company in order to move his work on polyamine analogues as therapeutic agents to the clinic. This work is now ongoing at Progen and as part of a number of academic collaborations. **Robert A. Casero, Jr** is a Professor of Oncology in the Johns Hopkins University School of Medicine. As a molecular pharmacologist he has spent most of the last 30 years studying the role of polyamines in normal and tumour cell growth, and devising strategies to target polyamine function and metabolism for therapeutic benefit. His laboratory was responsible for cloning several genes involved in human polyamine catabolism; genes whose expression is thought to affect cellular responses to specific polyamine analogues. Additionally, his laboratory participated in the discovery of LSD1 (lysine-specific demethylase 1). Dr Casero is also a scientific advisor to Progen Pharmaceuticals.

Nathaniel S. Rial is a graduate of The University of Arizona, College of Medicine. He also defended his dissertation at the University of Arizona, Tucson, AZ, U.S.A. in the Cancer Biology Programme. He is undertaking his residency at the University of Arizona in Internal Medicine. He is currently enrolled in the University of Arizona Mel and Enid Zuckerman College of Public Health. **Frank L. Meyskens, Jr** attended medical school at UCSF (University of California at San Francisco), then served his internship and residency at UCSF-Moffet Hospital. His haematology and oncology fellowship was at the National Cancer Institute within the National Institutes of Health campus. He is an investigator in trials focusing on therapy and chemoprevention in a variety of cancers, including melanoma and oral, colon, prostate and cervical cancer. He is currently Professor of Medicine and Biological Chemistry and Director of the Chao Family Comprehensive Cancer Center, and Associate Vice Chancellor of Health Sciences at the College of Health Sciences University of California, Irvine. **Eugene W. Gerner** began his graduate career in 1971 as The War on Cancer was signed into policy in the United States. He was influenced by Professor Alfred Knudson's lectures at the University of Texas embarking on a career in Cancer Biology. He was the founding Chair of the Cancer Biology Program at the University of Arizona, Tucson, AZ, U.S.A., and founding Director of the Gastrointestinal Cancer Program at the Arizona Cancer Center. He is Principal Investigator of a SPORE (Specialized Programs of Research Excellence) in gastrointestinal cancer. He is currently a Professor of Cell Biology and Anatomy and Biochemistry and Molecular Biophysics.

Leena Alhonen is a Professor of Animal Biotechnology and is responsible for the operation of the transgene unit at the University of Kuopio, Kuopio, Finland. She, together with the late Professor Juhani Jänne, started the pioneering work aimed at generation of transgenic rodents with altered polyamine homoeostasis to be used as tools to study the physiological roles of polyamines. The laboratory has produced more than 80 founder animals or rodent lines carrying polyamine metabolism-related transgenes. **Anne Uimari** is a senior researcher who participates in the molecular characterization of the polyamine metabolic genes and enzymes, and in the production and characterization of transgenic mice. She is also involved in the research studying the diseases connected to altered polyamine metabolism. **Marko Pietilä** works as a senior researcher in Leena Alhonen's group. He has specialized in the production of transgenic animal models. He has been closely involved on the production and characterization of both SSAT (spermidine/spermine N^1-acetyltransferase) transgenic and knockout animals. His special interests in the polyamine field are the roles of polyamines in the physiology of the skin and energy metabolism in adipose tissue. As a follow-on to his postdoctoral studies in the laboratory of Professor Coffino at UCSF (University of California at San Francisco), San Francisco, CA, U.S.A., he has started projects to assess the roles of antizyme and antizyme inhibitor in cancer cell growth. **Mervi T. Hyvönen** is a postdoctoral researcher in the animal biotechnology group. She has been investigating a variety of polyamine-related topics, such as the role of polyamines in acute pancreatitis, the regulation of SSAT expression and the role of eIF5A (eukaryotic translation initiation factor 5A) in polyamine depletion-induced cell-growth arrest. Her current studies aim to characterize the biological properties of several novel methylated polyamine analogues in order to develop research tools and potential therapeutics. **Eija Pirinen** is completing her Ph.D. studies in the area of molecular medicine at the University of Kuopio. Her Ph.D. studies revealed that the polyamine cycle can be considered as a futile cycle, and that activated polyamine catabolism regulates glucose and energy metabolism in mice. Her main research interest is energy metabolism in Type 2 diabetes, obesity and cardiac dysfunction. **Tuomo A. Keinänen** is a docent of chemical biology at the A.I. Virtanen Institute for Molecular Sciences, University of Kuopio. He is responsible for the analytical method development and chemical research tool development in polyamine research in the animal biotechnology group led by Professor Leena Alhonen.

Abbreviations

8-methyl-MMTA	5′-deoxy-5′-dimethylsulfonio-8-methyladenosine
AbeAdo	(5′-{[(Z)-4-amino-2-butenyl]methylamino}-5′-deoxyadenosine
AdoMet	*S*-adenosylmethionine
AdoMetDC	*S*-adenosylmethionine decarboxylase
AMA	*S*-(5′-deoxy-5′-adenosyl) methylthioethylhydroxylamine
ANXA1	annexin A1
APAO	N^1-acetylpolyamine oxidase
APC	adenomatous polyposis coli
Az	antizyme
AzI	antizyme inhibitor
CDK	cyclin-dependent kinase
CHO	Chinese hamster ovary
COX	cyclo-oxygenase
CRC	colorectal cancer
CV	cardiovascular
DAC	5-aza-2′-deoxycytidine
dcAdoMet	decarboxylated *S*-adenosylmethionine
DENSPM	diethylnorspermine
DFMO	difluoromethylornithine
EGF	epidermal growth factor
FAP	familial adenomatous polyposis
Genz-644131	5′-{[(Z)-4-amino-2-butenyl]methylamino}-5′-deoxy-8-methyladenosine
GSK-3β	glycogen synthase-3β
HAT	histone acetyl transferase
H3K4me	mono-methylated H3K4
H3K4me2	dimethylated H3K4
H3K4me3	tri-methylated H3K4
HDAC	histone deacetylase
IGF-1	insulin-like growth factor-1
Jmj C	Jumonji C
LSD1	lysine-specific demethylase 1
MAPK	mitogen-activated phosphate kinase
MAO	monoamine oxidase
MAOEA	5′-deoxy-5′-[(2-aminooxyethyl) methylamino]adenosine

MEK	MAPK (mitogen-activated protein kinase)/ERK (extracellular-signal-regulated kinase) kinase
MGBG	methylglyoxal bis(guanylhydrazone)
MHZPA	5′-deoxy-5′-[(3-hydrazinopropyl) methylamino]adenosine
MS-275	*N*-(2-aminophenyl)-4-[*N*-(pyridin-3-ylmethoxycarbonyl)-aminomethyl]benzamide
MTT	3-(4,5-dimethylthiazol-2-yl)-2,5-diphenyl-2*H*-tetrazolium bromide
NF-κB	nuclear factor κB
NMDA	*N*-methyl-D-aspartate
NO	nitric oxide
NOS-2	inducible nitric oxide synthase
NSAIDs	non-steroidal anti-inflammatory drugs
ODC	ornithine decarboxylase
ORF	open reading frame
PABA	polyaminobenzamide
PAHA	polyaminohydroxamic acid
PAO	polyamine oxidase
PDGF	platelet-derived growth factor
PGC-1α	PPARγ (peroxisome-proliferator-activated receptor γ) co-activator 1α
PGE_2	prostaglandin E_2
PI3K	phosphoinositide 3-kinase
PLP	pyridoxal phosphate
pRB	retinoblastoma protein
RNAi	RNA interference
R-point	restriction point
ROS	reactive oxygen species
SAHA	suberoylanilide hydroxamic acid
SAM486A	4-amidinoindan-1-one-2′-amidinohydrazone
SAT1	spermidine/spermine acetyltransferase (see also SSAT)
SFRP	secreted frizzle-related protein
SMO	spermine oxidase
SNP	single nucleotide polymorphism
SpdS	spermidine synthase
SpmS	spermine synthase
SSAT	spermidine/spermine N^1-acetyltransferase
SWIRM	Swi3p/Rsc8c/Moira
TCF/LEF	T-cell factor/lymphoid-enhancing factor
TRAMP	transgenic prostate adenocarcinoma model
uORF	upstream open reading frame
UTR	untranslated region

Essays Biochem. (2009) **46**, 1–9; doi:10.1042/BSE0460001

1

The polyamines: past, present and future

Heather M. Wallace[1]

Division of Applied Medicine, University of Aberdeen, Polwarth Building, Foresterhill, Aberdeen AB25 2ZD, Scotland, U.K.

The polyamines, spermidine and spermine, were first discovered in 1678 by Antonie van Leeuwenhoek. In the early part of the 20th century their structure was determined (Figure 1) and their pathway of biosynthesis established [1]. The polyamines are essential elements of cells from all species. They are required for optimum cell growth, and cells where polyamine production has been prevented by mutation, or blocked by inhibitors, require exogenous provision of at least one polyamine for continued survival. Despite this critical function, the polyamines have not attracted as much attention as they deserve in the wider field of biochemistry and cell biology. They are rarely mentioned in standard textbooks, despite over 75000 research papers having been written on the subject since 1900, and more than half (54%) were published after 1990 (A.E. Pegg, personal communication). There have been a number of books dedicated to the polyamines published and "*The Guide to the Polyamines*" by Seymour Cohen [2] deserves mention as a work of outstanding scholarship describing "everything you ever wanted to know about the polyamines" in exquisite detail. The current volume of *Essays in Biochemistry* has a much humbler aim: to introduce the polyamines to interested researchers and students, and to describe how they are associated with, and might be utilized as a target for intervention in major diseases such as cancer.

[1]*To whom correspondence should be addressed (email h.m.wallace@abdn.ac.uk).*

Putrescine
$H_3N^+(CH_2)_4N^+H_3$

Spermidine
$H_3N^+(CH_2)_4N^+H_2(CH_2)_3N^+H_3$

Spermine
$H_3N^+(CH_2)_3H_2N^+(CH_2)_4N^+H_2(CH_2)_3N^+H_3$

Figure 1. Structure of the polyamines

One of the main drivers in the quest to understand the function of the polyamines in mammalian cells has been their role in the growth and maintenance of cancer cells. Cancer is the second most common cause of morbidity and mortality world-wide, with millions of new cases of the common cancers, such as breast, colorectal, prostate and lung, diagnosed every year. More critically, the success rate in terms of cancer treatment varies enormously depending on the tumour type. The 5-year survival, the measure of success of therapy, is 80–90% for breast cancer, 45–55% for colorectal cancer and only approx. 5–15% for lung cancer. So clearly there is still a long way to go, and new drugs and treatments are required, particularly as the population ages. The polyamine pathway, because of the established link between polyamine concentrations in cells and cancer cell growth, has been identified as a goal for the development of antiproliferative agents.

The polyamine metabolic pathways have been recognized as a target for therapeutic intervention for four main reasons: (i) polyamines are required for cell growth; (ii) polyamine concentrations are increased significantly in cancer cells and tissues; (iii) ODC (ornithine decarboxylase), the first enzyme in the pathway, is increased in cancers and is an oncogene; and (iv) preventing polyamine biosynthesis prevents the growth of cells.

Another huge advantage is that polyamine metabolism is essential for all cancers, thus there is potential for treatment of multiple forms of the disease. In addition, more recent studies have suggested that inhibition of polyamine production may well be a novel and relatively non-toxic strategy for chemoprevention of cancer (see Chapter 8).

Polyamine metabolism is intriguing with a number of fascinating individual regulatory features which, if not unique, are at least uncommon in biology. This includes the rapid turnover of ODC, of the order of 10 min in mammalian cells, the +1 frameshifting that leads to translation of antizyme which targets ODC for degradation and the 'super-inducibility' of SSAT (spermidine/spermine N^1-acetyltransferase), all of which will be addressed in the present volume.

Spermidine, spermine and their diamine precursor, putrescine, are positively charged molecules at physiological pH (Figure 1). Spermidine and spermine interact electrostatically with DNA, but their charge is spread over the whole molecule rather than a point charge as found in mono- and di-valent cations making them subtly different in terms of charge. Both polyamines induce condensation and conformational changes in DNA and to some extent RNA [1]. It has been suggested that the polyamines are simply 'super' cations, but the question then arises: would a seven enzyme intracellular pathway, as well as uptake and export transport systems, evolve simply to provide a tri- or tetra-valent cation? Logic would dictate that this is not the case and that the polyamines and, indeed, their metabolism, have a more extensive role within the cell than being only polycations.

Classically, biosynthesis begins with ornithine and methionine, but recent evidence indicates a role for arginase in mammalian cells [1]. The major source of polyamines in the majority of mammalian cells is via *de novo* biosynthesis with diet playing a significant, but lesser, role. The smallest contribution to intracellular polyamine pools is made by the gut microflora (Figure 2).

Ornithine is decarboxylated by ODC to putrescine (Figure 3). Although strictly speaking not a polyamine, putrescine is usually considered along with spermidine and spermine as part of the polyamine family. Two aminopropyl groups are then added consecutively to putrescine to form spermidine, and to spermidine to form spermine. The aminopropyl groups are provided by decarboxylated *S*-adenosylmethionine which is itself produced from *S*-adenosylmethionine by the enzyme SAMDC (*S*-adenosylmethionine decarboxylase; also referred to as AdoMetDC in the present volume). The by-product of this reaction is 5′-methylthioadenosine which is recycled back to adenosine for further use. SAMDC is the subject of one of the chapters of this volume and will not be discussed further here (Chapter 3). The enzymes producing spermidine and spermine are synthases which are constitutive enzymes with little inducibility. This is in contrast with both ODC and SAMDC, which are readily induced by a range of agents. The ready inducibility of ODC by a variety of growth-promoting stimuli was one of the early indications that ODC was an important player in cell proliferation. ODC also has a fast

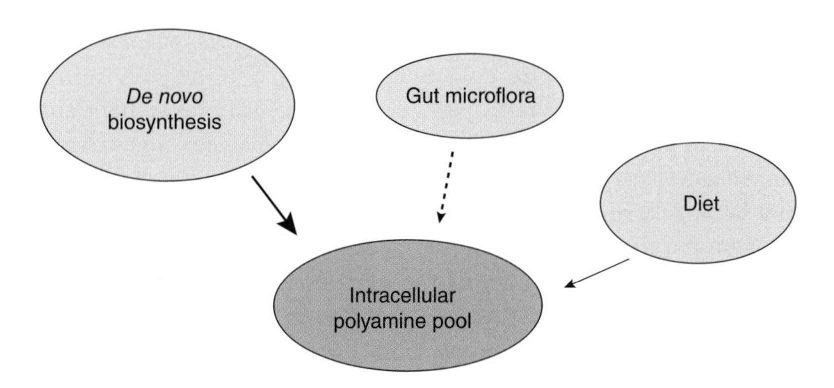

Figure 2. Sources of polyamines in mammalian cells

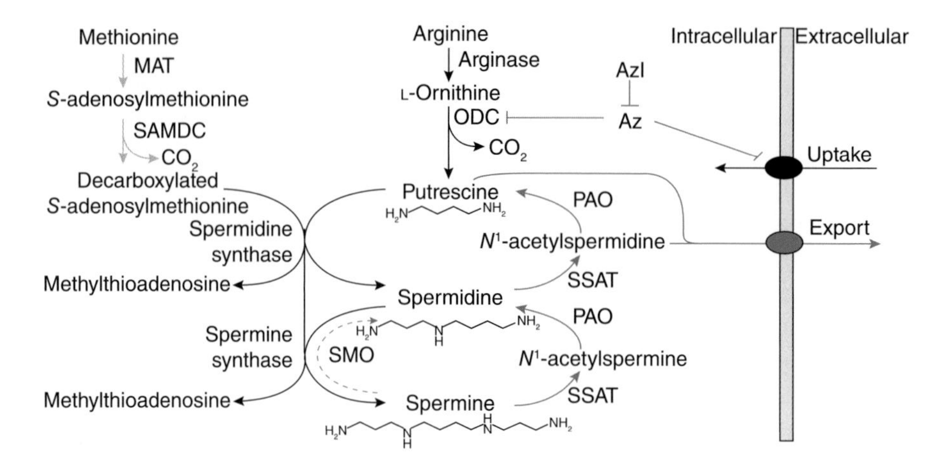

Figure 3. Polyamine pathways
Az, antizyme; AzI, antizyme inhibitor; MAT, methionine adenosyltransferase; PAO, polyamine oxidase. Reproduced with permission, from Wallace, H.W., Fraser, A.V. and Hughes, A. (2003) *Biochem. J.* **376**, 1–14. © the Biochemical Society.

turnover, of the order of 10 min, very fast for a mammalian enzyme. The regulation of ODC turnover by antizyme is also discussed later in this volume (Chapters 2 and 4).

Until quite recently polyamine biosynthesis was considered irreversible, a 'one-way street' from ornithine to spermine. Recycling of the polyamines does occur, but via a separate retroconversion pathway of acetylation catalysed by SSAT followed by oxidation catalysed by APAO (N^1-acetylpolyamine oxidase) [1]. The main reason for this two-step retroconversion seems to be to produce acetylated polyamine derivatives for export from the cell, providing a means of depleting intracellular polyamine concentrations [3]. N^1-Acetylspermidine and putrescine are the major polyamines exported from cells. In the early 2000s, Casero's group in Baltimore cloned another oxidase, SMO (spermine oxidase), which converts spermine directly into spermidine without the need for acetylation [4]. Bearing in mind that the acetylpolyamines are for export, it is logical that SMO is the enzyme used for recycling the polyamines, whereas the SSAT and APAO pathway is used for polyamine depletion.

In addition to the metabolic pathway there are also polyamine transporters: an inward, uptake transporter and an outward exporter. The transporters in mammalian cells are the last big challenge in polyamine metabolism as these have not been isolated or cloned from these cells. Evidence from yeast and bacteria suggest that there will be multiple transporters [5] and the current thinking is that many cells have at least two separate transporters, one for putrescine (and possibly other diamines) and one for spermidine and spermine. While research continues in this area, work from my own laboratory using a competitive, polyamine-uptake inhibitor suggests that, in human cancer cells, the importer and exporter are separate systems (H. Wallace, unpublished work).

As a result of the positive correlation between polyamines and cell growth it is perhaps not surprising that there is also a strong link between the polyamine content of cancer cells and their rate of growth. Indeed human cancer cells contain significantly more polyamine than the equivalent normal cells or tissue. For example, breast cancer cells have 4–6 times the polyamine content of normal breast cells, and colon cancer cells have 3–4 times the polyamine content of normal colonic cells [6,7]. Several groups had observed that patients with a variety of different cancers all showed increased urinary excretion of polyamines [8,9]. ODC activity was also increased in cancer cells, and so all of this was suggestive of a causal relationship and led many groups to look for targets in polyamine metabolism that would prevent polyamine biosynthesis and then perhaps tumour growth. The most obvious targets were the decarboxylase enzymes, ODC and SAMDC (Figure 3). A number of single enzyme inhibitors were designed and synthesized [10,11] with the hope of finding a new and effective anticancer drug [12].

The most successful inhibitor of this class is without doubt the irreversible suicide inhibitor of ODC, DFMO (α-difluoromethylornithine) [13]. This agent has stood the test of time well; having been synthesized in the late 1970s it has gone through several reincarnations. As a single-agent anticancer drug in man, DFMO was a disappointment. Rationally designed to inhibit a key pathway required for cancer cell growth it was hugely successful *in vitro* [13], but *in vivo* it failed to produce toxicity in cancer cells, and the compensatory increases in uptake of exogenous polyamines by cells treated with DFMO resulted in little effect on organ infiltration and overall survival [14]. However, as a single agent in parasitic disease (marketed as elfornithine) it is a great success producing cures for several trypanosomal infections (for a review, see [15]). Unfortunately, success in treating diseases in the developing world attracts little interest from drug companies, so although used and useful, DFMO is not a first-line drug in treatment of trypanasomiasis. It has found a small niche market as a hair-removing cream in a commercial product that uses 11% DFMO (www.vanqia.com). Recently, DFMO has emerged again, not as a chemotherapeutic agent, but as a strong candidate as a chemopreventative agent, the opportunities for which are discussed in Chapter 8.

Although DFMO was not useful as a single agent against cancer, it did provide a useful ‘proof-of-concept’ that inhibiting polyamine biosynthesis can prevent cancer cell growth, albeit in the absence of an external polyamine source. DFMO also showed that more extensive polyamine depletion is required. Decreases in all three polyamines are needed to prevent cancer cell growth, not just loss of putrescine and spermidine as occurs with DFMO. Finally, this research indicated that compensatory pathways need to be blocked as well to prevent attenuation of the initial inhibitory effects of the agent. A specific polyamine-uptake inhibitor would make DFMO an attractive antiproliferative drug.

Armed with the knowledge provided by nearly 20 years experience with DFMO, several groups developed polyamine analogues based on the following:

(i) analogues look like polyamines so they compete with the natural amines for transport (this may prevent the compensatory uptake seen with DFMO); (ii) analogues will result in negative-feedback inhibition of ODC and biosynthesis; and (iii) analogues are sufficiently different not to be able to mimic the function of the natural polyamines in cells (e.g. DNA condensation).

Three generations of polyamine analogues have now been synthesized from the simple symmetrically substituted bis(ethyl)spermidine and spermine compounds produced by Bergeron in Florida [17] through the unsymmetrically substituted mainly spermine analogues from Woster's group in Detroit [18], to the more complicated conformationally restricted and oligoamines synthesized originally by Frydmann and now produced by Progen [19]. These multiple analogues have had mixed success therapeutically, but have provided great tools for scientific research. The most interesting feature of treatment of cells with analogues is that many of them superinduce SSAT, which results in a rapid polyamine depletion from cells via the export transporter. This was an unexpected result, but since the natural polyamines induce SSAT, then perhaps it is not so surprising. Chapters 6 and 7 will discuss the development of these and a class of newer agents in more detail.

Cancer cells are effectively cells without barriers or checkpoints regulating their growth, which generally means a loss of regulation over the cell cycle. One of the early observations in cell-cycle studies showed that polyamines showed a biphasic response during the cell cycle [20], with ODC increasing in the G_1 phase and then again in G_2 with the subsequent increases in polyamine content. More recent work has linked these changes to interaction with oncogenes, cyclins and CDKs (cyclin-dependent kinases) which drive the cell cycle, and this will be discussed in more detail in Chapter 5.

Many cancers have up-regulated oncogenes, for example k-ras in colorectal cancer, and these up-regulated genes contribute to the overall disease and the potential for treatment. One set of key observations in the polyamines and cancer field is that showing the link between c-myc and ODC. The *ODC* gene contains a conserved repeat of the c-myc-binding site (CACGTG; an 'E-box' motif) suggesting that c-myc can regulate ODC at the level of transcription [21]. ODC *per se* has also been shown to be able to transform cells, suggesting that ODC is itself an oncogene [22].

Thus the weight of evidence linking polyamines and cancer looks very positive, with increased polyamine content or enzyme activity promoting the transformed phenotype. But what about the alternative: a link between polyamines and cell death? For many years it was hypothesized that any agent that caused cell death would by necessity have to induce loss of polyamines from cells because they were essentially growth promoting and therefore no longer required [1,10,23]. Early work showed that export of polyamines was the 'terminal' means of polyamine depletion, removing them from the cell and not recycling them [24]. This export process involved both metabolism and transport. Metabolism here is acetylation and oxidation, which provides the substrates for the outward transporter. This being the case, induction of cell death and also growth arrest should induce SSAT and APAO and increase the transport activity. The effect of the analogues on these processes has already been discussed, but other cytotoxic agents also induce these changes, including etoposide [25]. Thus induction, particularly of SSAT, may be considered a more general response to cellular stress leading to growth inhibition. The most common type of cell death induced by anticancer drugs is apoptosis, and polyamines have been found to both induce and inhibit apoptosis in mammalian cells. The bifunctional regulation of this key regulatory process remains to be fully understood. Another approach that has been taken in humans is to use a low-polyamine diet (a completely polyamine-free diet is not possible, practically). This work has been pioneered by the group of Moulinoux in Rennes where they have achieved some success in refractory cancer, particularly hormone refractory prostate cancer [26,27]. The original idea was to combine a low-polyamine diet with DFMO, an inhibitor of gut microflora metabolism (metronidazole) and an inhibitor of APAO (MDL 72,537) [28]. For various reasons this protocol has been cut down over the years, and with simply a low-polyamine diet alone improvements have been observed in patients. Perhaps the most interesting observations from these studies are the analgesic effects of the low-polyamine diet [26]. This may also be an exciting area for further development in the field.

An alternative approach to understanding polyamine and polyamine enzyme function in cells is to use transgenic technology to knock-out, knock-in or knock-down the key enzymes in the pathway. The group from Kuopio has led the way in this effort producing a range of transgenic animals with altered polyamine biochemistry (see Chapter 9).

The excitement in the field continues with new approaches being investigated all the time as technology advances. Currently, the use of DFMO as a chemopreventative agent against colorectal cancer is making huge strides forward with clinical trials ongoing. Similarly, the idea of using polyamines as vectors to introduce toxic agents more selectively to cancer cells is an area of growing interest with great potential for drug delivery [29].

In summary, the polyamines have a long history, are being actively studied at present and are set up for a bright future in chemoprevention and drug delivery.

References

1. Wallace, H.M., Fraser, A.V. and Hughes, A (2003) A perspective of polyamine metabolism. Biochem. J. **376**, 1–14
2. Cohen, S.S. (1998) A Guide to the Polyamines, Oxford University Press, New York
3. Wallace, H.M., Nuttall, M.E. and Coleman, C.S. (1988) Polyamine recycling enzymes in human cancer cells. Adv. Exp. Med. Biol. **250**, 331–344
4. Wang, Y., Murray-Stewart, T., Devereux, W., Hacker, A., Frydman, B., Woster, P.M. and Casero, Jr, R.A. (2003) Properties of purified recombinant human polyamine oxidase PAOh1/SMO. Biophys. Biochem. Res. Commun. **304**, 605–611

5. Igarashi, K., and Kashiwagi, K. (2009) Modulation of cellular functions by polyamines. Int. J. Biochem. Cell Biol. doi:10.1016/j.biocel.2009.07.009
6. Kingsnorth, A.N., Wallace, H.M., Bundred, N.J. and Dixon, J.M.J. (1984) Polyamines in breast cancer. Br. J. Surg. **71**, 352–356
7. Kingsnorth, A.N., Lumsden, A.B. and Wallace, H.M. (1984) Polyamines in colorectal cancer. Br. J. Surg. **71**, 791–794
8. Russell, D.H., Levy, C.C., Schimpff, S.C. and Hawk, I.A (1971) Urinary polyamines in cancer patients. Cancer Res. **31**, 1555–1558
9. Russell, D.H. and Durie, B.G.M. (1978) Polyamines as Biochemical Markers of Normal and Malignant Growth: Progress in Cancer Research and Therapy Volume 8, Raven Press, New York
10. Seiler, N. (2003) Thirty years of polyamine-related approaches to cancer therapy. Retrospect and prospect. Part 2: Structural analogues and derivatives. Curr. Drug Targets **4**, 565–585
11. Stanek, J., Frei, J., Mett, H., Schneider, P. and Regenass, U. (1992) 2-Substituted,3-(aminooxy) prop-anamines as inhibitors of ornithine decarboxylase: synthesis and biological activity. J. Med. Chem. **35**, 1339–1344
12. Metcalf, B.W., Bey, P., Danzin, C., Jung, M.J., Casara, P. and Vevert, J.P. (1978) Catalytic irreversible inhibition of mammalian ornithine decarboxylase (EC 4.1.1.17) by substrate and product analogues. J. Am. Chem. Soc. **100**, 2551–2553
13. Meyskens, Jr, F.L. and Gerner, E.W. (1999) Development of difluoromethylornithine (DFMO) as a chemoprevention agent. Clin. Cancer Res. **5**, 945–951
14. Smart, L.M., MacLachlan, G., Wallace, H.M. and Thomson, A.W. (1989) Influence of cyclosporine A and α-difluoromethylornithine, an inhibitor of polyamine biosynthesis, on two rodent T-cell cancers *in vivo*. Int. J. Cancer **44**, 1069–1073
15. Delespaux, V. and de Koning, H.P. (2007) Drugs and drug resistance in African trypanasomiasis. Drug Resist. Updates **10**, 30–50
16. Reference deleted
17. Porter, C.W., Cavanaugh, P.F., Stolowich, N., Ganis, B., Kelly, E. and Bergeron, R.J. (1985) Biological properties of N^4 and N^1,N^8-spermidine derivatives in cultured L1210 leukaemia cells. Cancer Res. **45**, 2050–2057
18. Saab, N.H., West, E.E., Bieszk, N.C., Preuss, C.V., Mank, A.R., Casero, Jr, R.A. and Woster, P.M. (1993) Synthesis and evaluation of unsymmetrically substituted polyamine analogues as modulators of human spermidine/spermine-N^1-acetyltransferase (SSAT) and as potential antitumor agents. J. Med. Chem. **36**, 2998–3004
19. Ruiz, O., Alonso-Garrido, D.O., Buldain, G, and Frydman, R.B. (1986) Effect of N-alkyl and C-alkylputrescines on the activity of ornithine decarboxylase from rat liver and E. coli. Biochim. Biophys. Acta **873**, 543–561
20. Heby, O. (1981) Role of polyamines in the control of cell proliferation and differentiation. Differentiation **19**, 1–20
21. Packham, G. and Cleveland, J.L. (1994) Ornithine decarboxylase is a mediator of c-Myc induced apoptosis. Mol. Cell. Biol. **14**, 5741–5747
22. Auvinen, M., Paasinen, A., Andersson, L.C. and Holtta, E. (1992) Ornithine decarboxylase activity is critical for cell transformation. Nature **360**, 355–358
23. Wallace, H.M. and Niiranen, K. (2007) Polyamine analogues: an update. Amino Acids **33**, 261–265
24. Wallace, H.M. and Keir, H.M. (1981) Uptake and excretion of polyamines from baby hamster kidney cells (BHK-21/C13). The effect of serum on confluent cultures. Biochim. Biophys. Acta **676**, 25–30
25. Lindsay, G.S. and Wallace, H.M. (1999) Changes in polyamine catabolism in HL60 human promyelogenous leukaemic cells in response to etoposide-induced apoptosis. Biochem. J. **337**, 83–87
26. Cipolla, B.G., Havouis, R. and Moulinoux, J.-P. (2007) Polyamine contents in current foods: a basis for polyamine reduced diet and a study of its long term observance and tolerance in prostate carcinoma patients. Amino Acids **33**, 203–212

27. Cipolla, B.G., Guilli, F. and Moulinoux, J.-P. (2003) Polyamine reduced diet in metastatic hormone-refractory prostate cancer (HRPC) patients. Biochem. Soc. Trans. **31**, 384–387
28. Chamaillard, L., Catros-Quemener, V., Delcros, J.G., Bansard, J.Y,, Havouis, R., Desury, D., Genetet, N. and Moulinoux, J.-P. (1997) Polyamine deprivation prevents the development of tumour-induced immune suppression. Br. J. Cancer **76**, 365–370
29. Phanstiel, IV, O., Kaur, N. and Delcros, J.G. (2007) Structure–activity investigations of polyamine–anthracene conjugates and their uptake via the polyamine transporter. Amino Acids **33**, 305–313

Essays Biochem. (2009) **46**, 11–24; doi:10.1042/BSE0460002

2

Polyamine homoeostasis

Lo Persson[1]

Department of Experimental Medical Science, Lund University, S-22184 Lund, Sweden

Abstract

The polyamines are essential for a variety of functions in the mammalian cell. Although their specific effects have not been fully elucidated, it is clear that the cellular polyamines have to be kept within certain levels for normal cell function. Polyamine homoeostasis in mammalian cells is achieved by a complex network of regulatory mechanisms affecting synthesis and degradation, as well as membrane transport of polyamines. The two key enzymes in the polyamine biosynthetic pathway, ODC (ornithine decarboxylase) and AdoMetDC (*S*-adenosylmethionine decarboxylase), are strongly regulated by feedback mechanisms at several levels, including transcriptional, translational and post-translational. Some of these mechanisms have been shown to be truly unique and include upstream reading frames and ribosomal frameshifting, as well as ubiquitin-independent proteasomal degradation. SSAT (spermidine/spermine N^1-acetyltransferase), which is a crucial enzyme for degradation and efflux of polyamines, is also highly regulated by polyamines. A cellular excess of polyamines rapidly induces SSAT, resulting in increased degradation/efflux of the polyamines. The polyamines appear to induce both transcription and translation of the *SSAT* mRNA. However, the major part of the polyamine-induced increase in SSAT is caused by a marked stabilization of the enzyme against degradation by the 26S proteasome. In addition, active transport of extracellular polyamines into the cell contributes to cellular polyamine homoeostasis. Depletion of cellular polyamines rapidly induces an increased uptake of exogenous polyamines, whereas an excess of polyamines

[1]*To whom correspondence should be addressed (Lo.Persson@med.lu.se).*

down-regulates the polyamine transporter(s). However, the protein(s) involved in polyamine transport and the exact mechanisms by which the polyamines regulate the transporter(s) are not yet known.

Introduction

A number of anabolic processes in the mammalian cells are highly dependent on adequate levels of polyamines [1,2]. A decrease in cellular polyamine concentrations often results in a reduction of nucleic acid and protein syntheses, eventually giving rise to cell growth inhibition [3]. Induction of cell growth is usually associated with an increase in cellular synthesis and levels of polyamines; however, too high concentrations of the polyamines may be toxic to the cells, inducing cell death or apoptosis. Thus cellular polyamine concentrations are usually maintained within rather narrow limits. This is achieved by a careful balance of synthesis, degradation and uptake of the polyamines [2,4]. These processes are all highly regulated and the polyamines exert a strong feedback control by affecting all of these steps in the polyamine metabolic pathway. In the present chapter the control of polyamine homoeostasis in mammalian cells will be described in more detail.

Polyamine synthesis

The polyamine biosynthetic pathway includes four different enzymes, i.e. ODC (ornithine decarboxylase), AdoMetDC [AdoMet (*S*-adenosylmethionine) decarboxylase], spermidine synthase and spermine synthase (Figure 1). The rate-limiting step in biosynthesis of polyamines appears to be either the decarboxylation of ornithine or the decarboxylation of AdoMet. These metabolic steps, catalysed by ODC and AdoMetDC, are also the key regulatory steps in the polyamine biosynthetic pathway. Both ODC and AdoMetDC have extremely high turnover rates in the cell [5–7], whereas spermidine synthase and spermine synthase are considerably more stable [8]. Owing to the fast turnover of ODC and AdoMetDC, the cellular levels of the enzyme proteins and thus the corresponding enzyme activities are rapidly altered when the synthesis and/or degradation of the enzymes are changed. Both enzymes are subject to a strong feedback control by the polyamines (Figure 1). ODC and AdoMetDC are rapidly up-regulated when cells become depleted of their polyamine content. On the other hand, both enzymes are down-regulated when cells are exposed to an excess of polyamines.

ODC

ODC is one of the enzymes in the mammalian cell having the shortest biological half-life [5,6]. Usually the half-life of ODC varies between 15 and 60 min, but can be as short as a few minutes in some cases. The turnover of ODC has been shown to be highly affected by the cellular polyamine levels [5,6,9]. Provision of an excess of polyamines to the cells will rapidly result

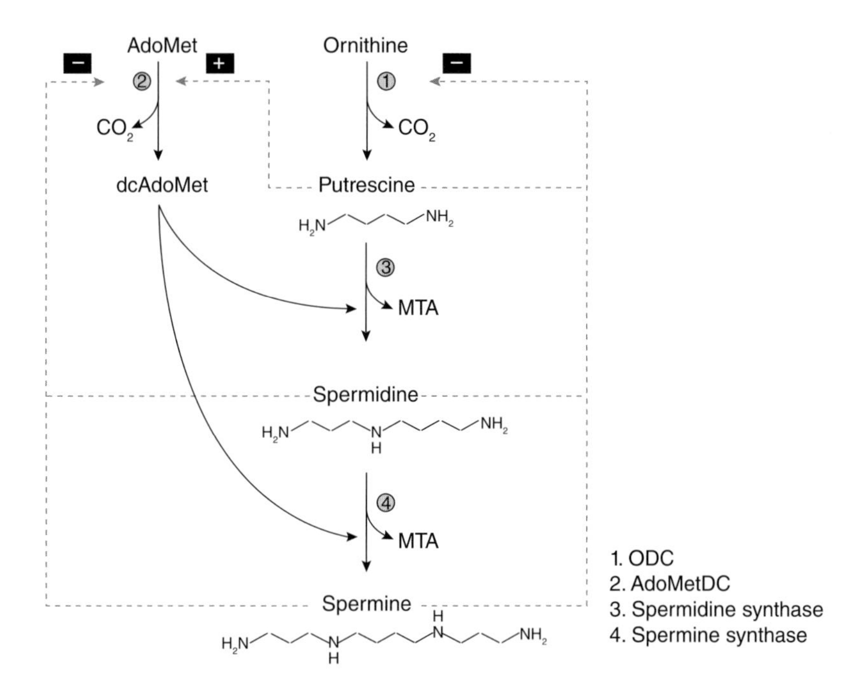

Figure 1. Feedback regulation of polyamine synthesis
The regulatory steps in polyamine biosynthesis are catalysed by ODC and AdoMetDC. Both of these enzymes have short half-lives and are regulated at various levels by the polyamines. Putrescine has a stimulatory effect on AdoMetDC processing and enzyme activity. dcAdoMet, decarboxylated AdoMet; MTA, methylthioadenosine.

in a faster degradation of the enzyme, reducing the half-life even further. On the other hand, a stabilization of ODC occurs when cells are depleted of their polyamines. All changes in the ODC degradation rate will rapidly result in a change in the level of ODC protein, and thus in the enzyme activity, due to the fast turnover of the enzyme. As a result of this, there is an inverse relationship between cellular polyamine content and ODC activity, which tends to stabilize polyamine levels in the cell. Significant progress has been made in the understanding of the mechanisms involved in the rapid degradation of ODC, as well as in how polyamines affect the turnover of the enzyme, revealing several unique processes. ODC is, like most other short-lived and regulatory proteins, degraded by the proteolytic complex called the 26S proteasome [5]. Usually, targeting proteins for degradation by the 26S proteasome involves covalent binding of multiple ubiquitin molecules to the protein. However, ODC is one of the very few proteins that are degraded by this proteolytic system without prior ubiquitination. In fact, ODC was the first example of a non-ubiquitinated protein being degraded by the 26S proteasome [5]. Instead, targeting of ODC for degradation by the 26S proteasome occurs by a binding of a specific protein, named Az (antizyme), to the enzyme (Figure 2) [5,6,9]. Az was discovered more than 30 years ago by Canellakis and colleagues [10]. They demonstrated that exposure of cells to high levels of putrescine resulted

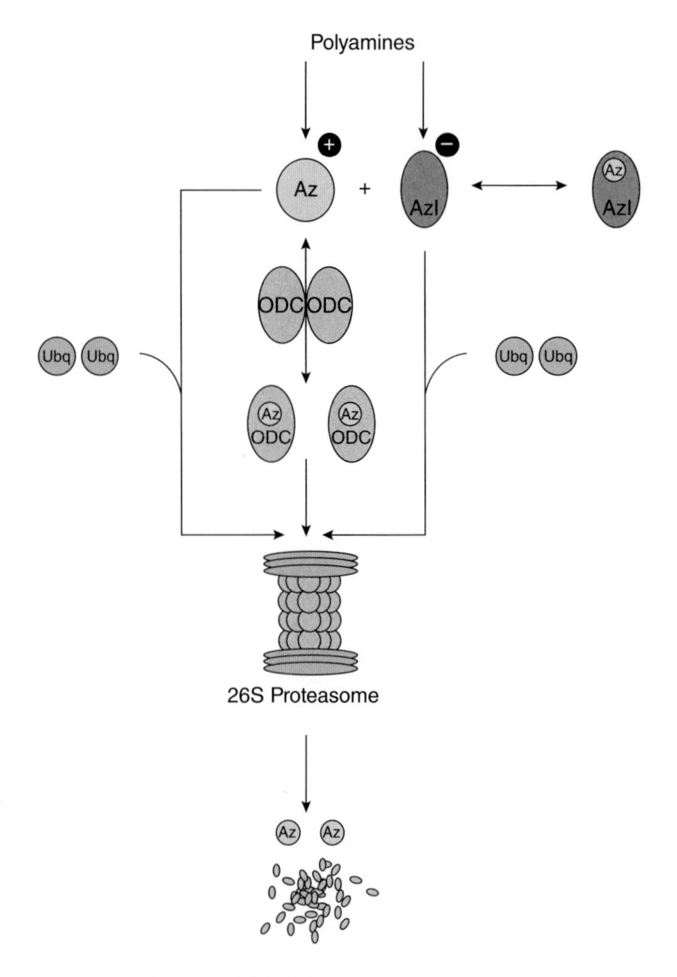

Figure 2. Feedback regulation of ODC degradation

The degradation of ODC by the 26S proteasome is induced by the binding of Az to the enzyme. Polyamines stimulate Az production by ribosomal frameshifting (see Figure 3). AzI binds Az with stronger affinity than ODC, reducing the amount of free Az (and thus ODC degradation). Polyamines inhibit *AzI* mRNA translation. Both Az and AzI are degraded by the 26S proteasome in an ubiquitin-dependent process. Ubq, ubiquitin.

in the induction of a unique short-lived protein, which strongly inhibited ODC activity. They also proposed the name Az for this group of inhibitory proteins. However, it was not until later that it was demonstrated that, besides inhibiting the ODC activity, Az induced the degradation of the ODC protein by the 26S proteasome [4–6,9]. ODC is functionally a dimer and the two active sites, which are formed at the dimer interface, contain amino acids from both subunits [4]. The binding of Az occurs at the monomer level and thus promotes the dissociation and inactivation of the dimer (Figure 2) [6].

The synthesis of Az is stimulated by polyamines. Thus an excess of polyamines down-regulates ODC by affecting the production of Az, which inhibits the enzyme as well as stimulates its degradation [5]. On the other hand, a depletion of cellular polyamine content results in a stabilization of ODC protein due to a decrease in Az production. Studies of the mechanism by which the polyamines affect the synthesis of Az revealed a truly unique regulatory mechanism. The cloning and sequencing of Az mRNA demonstrated that the message contains two major, partially overlapping, ORFs (open reading frames) and that the expression of the full-length Az required a ribosomal +1 frameshift (Figure 3) [5,6]. It was also shown that polyamines stimulated the synthesis of Az by promoting the +1 frameshifting [11]. Programmed ribosomal frameshifting is an extremely rare event and is mainly known from various viruses. So far, synthesis of Az is the only known example of a regulated frameshifting occurring in mammalian cells.

Truncations and mutations in the mammalian ODC have revealed that the C-terminal part of the enzyme is essential for the rapid degradation of the protein [12]. Interestingly, ODC from the parasite (*Trypanosoma brucei*) causing African sleeping sickness lacks this part of the protein and is a metabolically stable protein [13]. However, if *T. brucei* ODC is recombined with the C-terminal part of mammalian ODC the parasitic enzyme is transformed into a short half-life protein when expressed in mammalian cells [13]. The C-terminal part of mammalian ODC has been shown to be essential for the recognition by the proteasome and binding of Az to ODC is believed to affect the structure of the monomer in such a way that the C-terminal part is exposed, targeting the protein for degradation by the 26S proteasome [6].

The amount of Az available for ODC binding is not only regulated at the translational level. In addition, Az activity is regulated by a specific protein

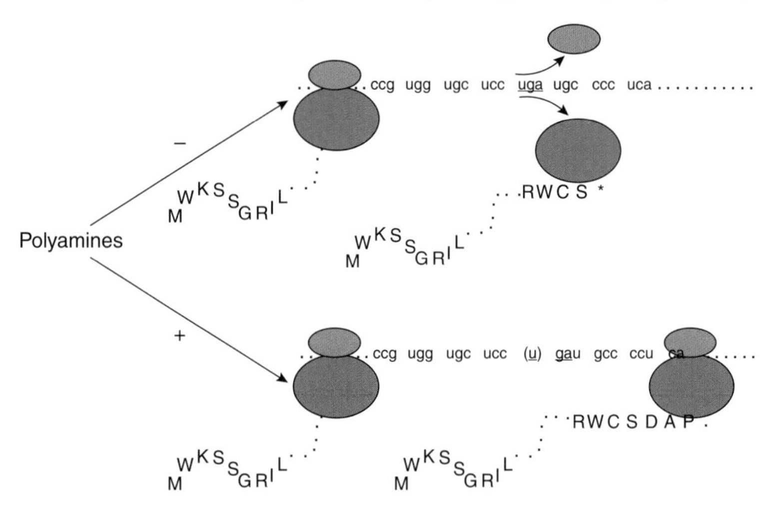

Figure 3. Ribosomal frameshifting in translation of *Az* mRNA
Synthesis of full-length Az requires translation of two partially overlapping ORFs in the *Az* mRNA. This is achieved by a +1 frameshift at the end of the first ORF. The frameshift is stimulated by polyamines. In the presence of low polyamine levels, translation will come to an end at the UGA stop codon (underlined) of the first ORF, giving rise to a truncated form of Az. However, in the presence of high polyamine levels a larger fraction of the ribosomes will continue translation of the second ORF after a +1 frameshift at the UGA stop codon, giving rise to full-length Az.

named AzI (Az inhibitor) (Figure 2) [4,5,9]. AzI is highly homologous with ODC without being enzymatically active (lacks essential amino acids). However, like ODC, AzI binds Az. The binding affinity between AzI and Az is stronger than that between ODC and Az. Thus, in the presence of AzI, less Az is available for binding to ODC, which results in a decreased degradation of the enzyme. Even though AzI most probably fulfils an important function in polyamine homoeostasis, not much is known about its regulation. To make everything even more complicated, three different forms of Az (Az1–Az3) and two different forms of AzI (AzI-1 and AzI-2) have been discovered [6,9]. However, it was recently demonstrated that *AzI* mRNA contains a small uORF (upstream ORF), which mediates polyamine-induced inhibition of the translation of the downstream main ORF [14]. Thus, in the presence of polyamines, AzI should be down-regulated, giving rise to a reduction of ODC due to degradation by the 26S proteasome. The importance of AzI was also recently demonstrated using knockout mice with a disrupted *Oaz1* (mouse AzI-1) gene [15]. Homozygous mutant mice died at the newborn stage with abnormal liver morphology. Further analysis of embryos of the homozygous mice indicated that the deletion of the *Oaz1* gene induced a degradation of ODC and a reduction of polyamine biosynthesis.

Both Az and AzI have a rapid turnover in the cell and are degraded by the 26S proteasome in an ubiquitin-dependent process [9]. Even though Az binding to ODC induced degradation of ODC by the proteasome, degradation of Az was unaffected by the interaction with ODC. On the other hand, binding of Az to AzI stabilized AzI against degradation by the proteasome.

In addition, to induce changes in ODC turnover rate, the polyamines appear to affect ODC synthesis. However, this feedback control of ODC synthesis is not correlated with any changes in *ODC* mRNA levels, indicating an effect on translation rather than on transcription (or stability) of the *ODC* mRNA [16]. Like many of the other mRNAs encoding growth-related proteins, mammalian *ODC* mRNA contains a long GC-rich 5′ UTR (untranslated region), which strongly inhibits the translation of the subsequent ORF. The expression of ODC has been shown to be stimulated by the initiation factor eIF-4E (eukaryotic initiation factor 4E). This initiation factor contains an RNA helicase activity, which is important for the melting of secondary structures of the 5′ UTR. Moreover, the 5′ UTR of ODC mRNA has been suggested to contain an IRES (internal ribosomal entry site), making cap-independent translation possible [17]. However, so far no evidence exists that the 5′ UTR of *ODC* mRNA is important for the polyamine-mediated control of ODC synthesis.

AdoMetDC

Synthesis of spermidine and spermine is achieved by the addition of an aminopropyl group to putrescine and spermidine respectively. These aminopropyl groups are derived from decarboxylated AdoMet, which is produced by AdoMetDC. Mammalian AdoMetDC belongs to the group of very few enzymes that uses pyruvate, instead of pyridoxal 5′-phosphate, as a cofactor [18]. The pyruvate is generated by an autocatalytical cleavage of the AdoMetDC protein after synthesis [19]. Mammalian AdoMetDC is produced as a pro-enzyme which is rapidly cleaved into two differently sized subunits, which both are part of the active heterotetrameric form of the enzyme [18]. The cleavage occurs at a serine residue (in the larger subunit), which then is converted into pyruvate. The processing of the AdoMetDC pro-enzyme has been shown to be stimulated by putrescine, which also has a direct stimulatory effect on the catalytic activity of the enzyme [18].

Mammalian AdoMetDC is also highly regulated and its turnover is very fast. Like mammalian ODC, AdoMetDC is degraded by the 26S proteasome [7]. However, unlike the degradation of ODC by the 26S proteasome, the degradation of AdoMetDC is dependent on ubiquitination of the protein. Interestingly, the degradation rate appears to be very dependent on the availability of the substrate AdoMet [7]. In the presence of low substrate levels, the degradation of AdoMetDC is decreased, whereas high substrate levels are believed to induce a more rapid degradation. The binding of AdoMet in the active site occasionally gives rise to transamination of an amino group from the substrate to the bound pyruvate, converting it into alanine and thus irreversibly inactivating the enzyme [7]. Although this occurs at a very low frequency, it may be substantial in the presence of high concentrations of the substrate. The transaminated form of AdoMetDC appears to be more rapidly degraded than the active form of AdoMetDC, explaining the substrate-mediated difference in degradation rate [7]. In addition, the turnover of AdoMetDC is affected by the cellular supply of spermidine and spermine in such a way that decreased polyamine levels give rise to a stabilization of the enzyme against degradation [18]. However, the mechanism by which the polyamines affect the degradation of AdoMetDC is still unknown.

The polyamines spermidine and spermine negatively regulate the synthesis of AdoMetDC [18]. Both polyamines are effective, but spermine appears to be a more potent regulator than spermidine. Part of this control appears to be achieved at the transcriptional level, since the *AdoMetDC* mRNA level is changed according to the polyamine status of the cells. Depletion of cellular spermidine and/or spermine induces an increase, whereas an excess of these polyamines results in a decrease in the *AdoMetDC* mRNA level. In addition to affecting the *AdoMetDC* mRNA steady-state level, spermidine and spermine induce changes in the efficiency by which the *AdoMetDC* mRNA is translated [18]. The efficiency is increased in situations of low polyamine levels and vice versa. This effect by the polyamines appears to be dependent on a short uORF of the *AdoMetDC* mRNA [20]. The uORF codes for the peptide MAGDIS and the production of this peptide seems to be essential for the polyamine-mediated effect on the translation of the AdoMetDC reading frame. Mutating the initiation codon of the uORF, as well as changing the nucleotide sequence coding for the C-terminus of the peptide, abolishes the polyamine-mediated regulation of *AdoMetDC* mRNA translation [20]. It has been shown that, during translation of the uORF, ribosomes pause close to the termination codon. Changing sequences coding for the C-terminus of the MAGDIS peptide reduces the pausing of the ribosomes making translation of the downstream reading frame more efficient. The time the ribosomes are stalled is dependent on the polyamines. The polyamines prolong the pausing period, which reduces downstream translation of *AdoMetDC* mRNA [20].

Polyamine degradation

Spermidine and spermine may be degraded to putrescine and spermidine respectively, in a two-step process usually referred to as 'the polyamine interconversion pathway' (Figure 4) [21]. The first step in this pathway is an acetylation of the N^1-nitrogen of the polyamine producing N^1-acetylspermidine or N^1-acetylspermine, catalysed by the enzyme SSAT (spermidine/spermine N^1-acetyltransferase). The acetylation is followed by an oxidation catalysed by a FAD-dependent peroxisomal polyamine oxidase. The preferred substrates of this polyamine oxidase are the N^1-acetylated derivatives of spermidine and spermine. Spermidine and spermine are poor substrates of this enzyme unless they are acetylated. The enzyme is nowadays referred to

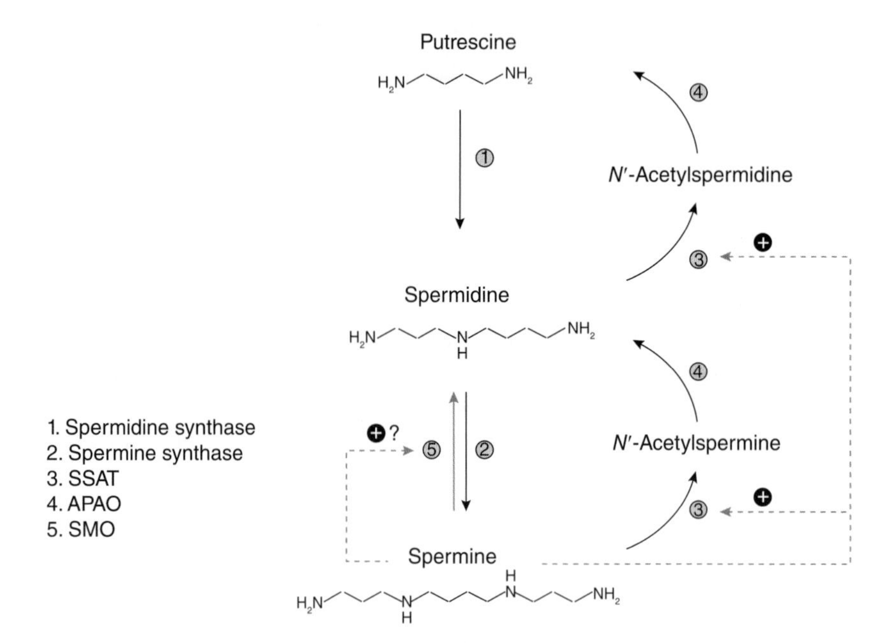

Figure 4. Feedback regulation of polyamine degradation
Spermidine and spermine may be converted into putrescine and spermidine respectively, in a two-step process, catalysed by SSAT and APAO. The rate-limiting step is catalysed by SSAT, which has a fast turnover and is induced by the polyamines. Spermine can also be converted into spermidine in a single step catalysed by SMO. SMO expression is induced by polyamine analogues, and most likely also by spermine.

as the APAO (N^1-acetylpolyamine oxidase) to distinguish it from the recently discovered SMO (spermine oxidase), which preferentially uses spermine as a substrate (see below). The results of the oxidation of N^1-acetylspermidine and N^1-acetylspermine by APAO are putrescine and spermidine respectively, as well as 3-acetamidopropanal and hydrogen peroxide [21]. Putrescine may be further oxidized by diamine oxidase, whereas spermidine may undergo another round in the interconversion pathway. The acetylated derivatives of spermidine and spermine may also be excreted from the cells. The exact mechanisms by which the acetylated polyamines are excreted are still not clear. However, recently SLC3A2, which is part of a heterodimeric cationic amino acid transporter, was identified as being involved in the export of especially putrescine, but also of acetylated spermidine [22]. The SLC3A2 protein was shown to be co-localized with SSAT on the plasma membrane, and both proteins were co-immunoprecipitated using antibodies against either SLC3A2 or SSAT, suggesting a functional interaction [22]. Interestingly, SLC3A2-dependent export of putrescine appeared to be associated with uptake of arginine, indicating an exchange reaction. Arginine, being a substrate for ornithine production, is essential for polyamine biosynthesis. However, whether this efflux system plays a physiological role in cellular polyamine homoeostasis or not remains to be established.

SSAT catalyses the rate-limiting step in the polyamine interconversion pathway [23]. The activity of APAO normally greatly exceeds that of the acetylase. Like ODC and AdoMetDC, SSAT has a very fast turnover with a half-life as short as 15 min, whereas APAO is a stable enzyme [23]. Thus SSAT may respond quickly to changes in the synthesis or degradation of the enzyme. The rapid turnover of SSAT is mediated by the 26S proteasome and is dependent on ubiquitination of the protein [24]. SSAT is strongly induced by a variety of stimuli, including various toxins and hormones. However, the enzyme is also induced by polyamines (Figure 4) and some of their analogues, such as N^1,N^{12}-bis(ethyl)spermine and N^1,N^{12}-bis(ethyl)norspermine [24]. Thus, similar to ODC and AdoMetDC, SSAT plays an important role in polyamine homoeostasis. A minor part of the increase in SSAT activity observed after a polyamine load is explained by an increase in the level of SSAT mRNA, indicating a transcriptional effect. A putative polyamine-responsive element has been identified in the promoter of the *SSAT* gene [25]. Part of the increase in SSAT expression may be due to an increase in the efficiency by which the *SSAT* mRNA is translated. However, the major part of the polyamine-induced increase in SSAT appears to be caused by a marked stabilization of the enzyme against degradation. SSAT degradation has been shown to be strongly inhibited by the binding of polyamines or N^1,N^{12}-bis(ethyl)spermine to the enzyme [24]. The rapid turnover of SSAT, like that of ODC, is dependent on the C-terminal region of the protein. Truncation or mutations in this region completely abolish the rapid degradation of SSAT [24]. It is conceivable that the binding of polyamines or their analogues to SSAT induce a conformational change in the enzyme protein that prevents the exposure of the C-terminal region to the proteasome.

In addition to being a substrate for the polyamine interconversion pathway, spermine may be degraded to spermidine, 3-aminopropanal and hydrogen peroxide in a single step catalysed by the enzyme SMO (Figure 4) [26,27]. Like APAO, SMO is a FAD-dependent oxidase but, unlike the former enzyme, SMO has a high specificity for spermine as a substrate. In great contrast with APAO, SMO is highly induced by several of the polyamine analogues, indicating a role in polyamine homoeostasis [26]. However, the regulation of SMO appears to be mainly at the level of mRNA transcription/stability, rather than translational/post-translational [28].

Polyamine uptake

In addition to regulating polyamine levels by synthesis, degradation and efflux, cells are equipped with an active transport system for the uptake of polyamines [29]. Large amounts of polyamines are found in the food we eat. Bacteria in the intestinal system produce and excrete considerable quantities of polyamines. It is conceivable that a large fraction of these exogenous polyamines are absorbed from the intestines and later taken up and used by cells in the body. However, to what extent cells rely upon endogenous compared with exogenous polyamines is not yet clear. Nevertheless, in situations when the polyamine biosynthetic machinery is insufficient, cells would certainly be more dependent on extracellularly derived polyamines. Results from experiments using cells deficient in polyamine transport indicate that the transport system may also be involved in the salvage of polyamines normally excreted from cells [30].

The exact mechanisms and the proteins involved in polyamine transport are still not identified. Whether there are individual transport systems for the various polyamines or only a single transporter, capable of transporting all of the polyamines, is not clear. Results obtained indicate that uptake of polyamines by mammalian cells, at least partly, occurs by a mechanism involving cell-surface heparin sulfate proteoglycans and endocytosis [31]. The polysaccharides of the proteoglycans are negatively charged and may interact with the positively charged polyamines with affinities even stronger than the interaction between DNA and polyamines. Moreover, it was recently demonstrated that polyamine uptake in human colon cancer cells follows a dynamin-dependent and clathrin-independent endocytic route, which is negatively regulated by caveolin-1 [32]. However, in spite of large efforts, a mammalian transporter responsible for polyamine uptake has not yet been cloned. Nevertheless, a cell-surface protein, capable of transporting both putrescine and spermidine, has been cloned from the protozoan pathogen *Leishmania major* [33].

The polyamine transporter is an important part of the polyamine homoeostatic system in mammalian cells [29]. The activity of the polyamine transporter is partly regulated by cellular polyamine levels. Cellular depletion of polyamines results in a marked increase in the cellular uptake of exogenous polyamines. On the other hand, the polyamine transporter is down-regulated when cells are exposed to an excess of polyamines. This feedback regulation is dependent on protein synthesis and involves a protein with a very fast turnover [29]. Interestingly, it has been shown that Az, which is induced by an excess of polyamines (and has a fast turnover), besides regulating the degradation of ODC, also appears to negatively regulate the cellular polyamine transporter [29]. Cells in which Az is expressed to high levels exhibit a marked reduction in polyamine uptake. All three different forms of Az (Az1–Az3) have been shown to effectively down-regulate polyamine transport. However, the mechanism by which Az affects polyamine uptake is so far unknown.

It is conceivable that the therapeutic effect of inhibitors of ODC or AdoMetDC against proliferative disorders, such as cancer, may partly be neutralized by an increased cellular uptake of exogenous polyamines [34]. Any decrease in cellular polyamine levels will rapidly induce an elevated uptake of polyamines from the extracellular environment. Results from various model systems strongly indicate that the therapeutic effect of the ODC inhibitors could be dramatically improved if the uptake of exogenous polyamines can be blocked [34,35].

Conclusions

Cellular polyamine homoeostasis is achieved by effective feedback mechanisms at a multitude of levels, including synthesis, degradation and transport (Figure 5). At present, we understand some of these mechanisms in part, but most of the molecular details are still unknown. The polyamine metabolic pathway is a potential target for therapeutic agents against a variety of diseases,

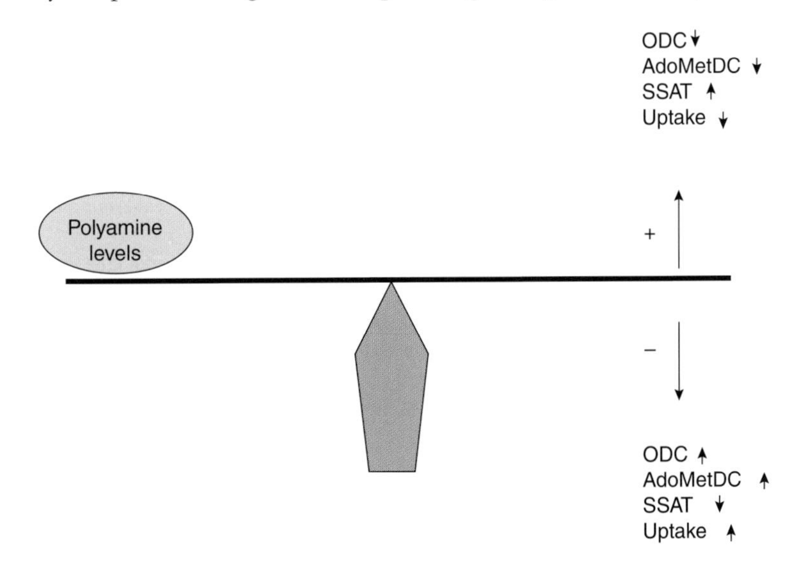

Figure 5. Regulation of polyamine homoeostasis
Polyamine homoeostasis is achieved by a balanced control of polyamine synthesis, degradation and uptake.

including cancer. However, the usefulness of these agents may depend on the understanding and control of the molecular mechanisms involved in the cellular control of polyamine homoeostasis. Moreover, such understanding may constitute the basis for future development of new and more effective drugs interfering with cellular polyamine homoeostasis.

Summary

- *Polyamine homoeostasis in mammalian cells is achieved by a complex network of regulatory mechanisms affecting synthesis, degradation and uptake of the polyamines.*
- *All key enzymes in polyamine biosynthesis and degradation have extremely short half-lives and are subject to strong feedback regulation by the polyamines at various levels.*
- *Some of the mechanisms involved in polyamine homoeostasis are truly unique, including, for example, upstream reading frames and ribosomal frameshifting, as well as ubiquitin-independent proteasomal degradation.*
- *The polyamine metabolic pathway is a potential target for therapeutic agents against a variety of diseases, and the understanding of the molecular mechanisms involved in cellular polyamine homoeostasis may form the basis for development of new and more effective drugs.*

References

1. Wallace, H.M., Fraser, A.V. and Hughes, A. (2003) A perspective of polyamine metabolism. Biochem. J. **376**, 1–14
2. Casero, Jr, R.A. and Marton, L.J. (2007) Targeting polyamine metabolism and function in cancer and other hyperproliferative diseases. Nat. Rev. Drug Discov. **6**, 373–390
3. Gerner, E.W. and Meyskens, Jr, F.L. (2004) Polyamines and cancer: old molecules, new understanding. Nat. Rev. Cancer **4**, 781–792
4. Pegg, A.E. (2006) Regulation of ornithine decarboxylase. J. Biol. Chem. **281**, 14529–14532
5. Hayashi, S., Murakami, Y. and Matsufuji, S. (1996) Ornithine decarboxylase antizyme: a novel type of regulatory protein. Trends Biochem. Sci. **21**, 27–30
6. Coffino, P. (2001) Regulation of cellular polyamines by antizyme. Nat. Rev. Mol. Cell Biol. **2**, 188–194
7. Yerlikaya, A. and Stanley, B.A. (2004) S-Adenosylmethionine decarboxylase degradation by the 26S proteasome is accelerated by substrate-mediated transamination. J. Biol. Chem. **279**, 12469–12478
8. Ikeguchi, Y., Bewley, M.C. and Pegg, A.E. (2006) Aminopropyltransferases: function, structure and genetics. J. Biochem. **139**, 1–9
9. Kahana, C. (2009) Antizyme and antizyme inhibitor, a regulatory tango. Cell Mol. Life Sci., doi: 10.1007/s00018-009-0033-3
10. Heller, J.S., Fong, W.F. and Canellakis, E.S. (1976) Induction of a protein inhibitor to ornithine decarboxylase by the end products of its reaction. Proc. Natl. Acad. Sci. U.S.A. **73**, 1858–1862
11. Matsufuji, S., Matsufuji, T., Miyazaki, Y., Murakami, Y., Atkins, J.F., Gesteland, R.F. and Hayashi, S. (1995) Autoregulatory frameshifting in decoding mammalian ornithine decarboxylase antizyme. Cell **80**, 51–60

12. Ghoda, L., van Daalen Wetters, T., Macrae, M., Ascherman, D. and Coffino, P. (1989) Prevention of rapid intracellular degradation of ODC by a carboxyl-terminal truncation. Science **243**, 1493–1495
13. Ghoda, L., Phillips, M.A., Bass, K.E., Wang, C.C. and Coffino, P. (1990) Trypanosome ornithine decarboxylase is stable because it lacks sequences found in the carboxyl terminus of the mouse enzyme which target the latter for intracellular degradation. J. Biol. Chem. **265**, 11823–11826
14. Ivanov, I.P., Loughran, G. and Atkins, J.F. (2008) uORFs with unusual translational start codons autoregulate expression of eukaryotic ornithine decarboxylase homologs. Proc. Natl. Acad. Sci. U.S.A. **105**, 10079–10084
15. Tang, H., Ariki, K., Ohkido, M., Murakami, Y., Matsufuji, S., Li, Z. and Yamamura, K. (2009) Role of ornithine decarboxylase antizyme inhibitor *in vivo*. Genes Cells **14**, 79–87
16. Shantz, L.M. and Pegg, A.E. (1999) Translational regulation of ornithine decarboxylase and other enzymes of the polyamine pathway. Int. J. Biochem. Cell Biol. **31**, 107–122
17. Pyronnet, S., Pradayrol, L. and Sonenberg, N. (2000) A cell cycle-dependent internal ribosome entry site. Mol. Cell **5**, 607–616
18. Pegg, A.E., Xiong, H., Feith, D.J. and Shantz, L.M. (1998) S-Adenosylmethionine decarboxylase: structure, function and regulation by polyamines. Biochem. Soc. Trans. **26**, 580–586
19. Tolbert, W.D., Zhang, Y., Cottet, S.E., Bennett, E.M., Ekstrom, J.L., Pegg, A.E. and Ealick, S.E. (2003) Mechanism of human S-adenosylmethionine decarboxylase proenzyme processing as revealed by the structure of the S68A mutant. Biochemistry **42**, 2386–2395
20. Ruan, H.J., Shantz, L.M., Pegg, A.E. and Morris, D.R. (1996) The upstream open reading frame of the mRNA encoding S-adenosylmethionine decarboxylase is a polyamine-responsive translational control element. J. Biol. Chem. **271**, 29576–29582
21. Seiler, N. (2004) Catabolism of polyamines. Amino Acids **26**, 217–233
22. Uemura, T., Yerushalmi, H.F., Tsaprailis, G., Stringer, D.E., Pastorian, K.E., Hawel, III, L., Byus, C.V. and Gerner, E.W. (2008) Identification and characterization of a diamine exporter in colon epithelial cells. J. Biol. Chem. **283**, 26428–26435
23. Casero, Jr, R.A. and Pegg, A.E. (1993) Spermidine/spermine N1-acetyltransferase: the turning point in polyamine metabolism. FASEB J. **7**, 653–661
24. Coleman, C.S. and Pegg, A.E. (2001) Polyamine analogues inhibit the ubiquitination of spermidine/spermine N1-acetyltransferase and prevent its targeting to the proteasome for degradation. Biochem. J. **358**, 137–145
25. Wang, Y., Xiao, L., Thiagalingam, A., Nelkin, B.D. and Casero, Jr, R.A. (1998) The identification of a *cis*-element and a *trans*-acting factor involved in the response to polyamines and polyamine analogues in the regulation of the human spermidine/spermine N1-acetyltransferase gene transcription. J. Biol. Chem. **273**, 34623–34630
26. Wang, Y., Devereux, W., Woster, P.M., Stewart, T.M., Hacker, A. and Casero, Jr, R.A. (2001) Cloning and characterization of a human polyamine oxidase that is inducible by polyamine analogue exposure. Cancer Res. **61**, 5370–5373
27. Vujcic, S., Diegelman, P., Bacchi, C.J., Kramer, D.L. and Porter, C.W. (2002) Identification and characterization of a novel flavin-containing spermine oxidase of mammalian cell origin. Biochem. J. **367**, 665–675
28. Wang, Y., Hacker, A., Murray-Stewart, T., Fleischer, J.G., Woster, P.M. and Casero, Jr, R.A. (2005) Induction of human spermine oxidase SMO(PAOh1) is regulated at the levels of new mRNA synthesis, mRNA stabilization and newly synthesized protein. Biochem. J. **386**, 543–547
29. Mitchell, J.L., Thane, T.K., Sequeira, J.M. and Thokala, R. (2007) Unusual aspects of the polyamine transport system affect the design of strategies for use of polyamine analogues in chemotherapy. Biochem. Soc. Trans. **35**, 318–321
30. Byers, T.L. and Pegg, A.E. (1989) Properties and physiological function of the polyamine transport system. Am. J. Physiol. **257**, C545–C553
31. Welch, J.E., Bengtson, P., Svensson, K., Wittrup, A., Jenniskens, G.J., Ten Dam, G.B., Van Kuppevelt, T.H. and Belting, M. (2008) Single chain fragment anti-heparan sulfate antibody targets the polyamine transport system and attenuates polyamine-dependent cell proliferation. Int. J. Oncol. **32**, 749–756

32. Roy, U.K., Rial, N.S., Kachel, K.L. and Gerner, E.W. (2008) Activated K-RAS increases polyamine uptake in human colon cancer cells through modulation of caveolar endocytosis. Mol. Carcinog. **47**, 538–553
33. Hasne, M.P. and Ullman, B. (2005) Identification and characterization of a polyamine permease from the protozoan parasite *Leishmania major*. J. Biol. Chem. **280**, 15188–15194
34. Burns, M.R., Graminski, G.F., Weeks, R.S., Chen, Y. and O'Brien, T.G. (2009) Lipophilic lysine–spermine conjugates are potent polyamine transport inhibitors for use in combination with a polyamine biosynthesis inhibitor. J. Med. Chem. **52**, 1983–1993
35. Persson, L., Holm, I., Ask, A. and Heby, O. (1988) Curative effect of DL-2-difluoromethylornithine on mice bearing mutant L1210 leukemia cells deficient in polyamine uptake. Cancer Res. **48**, 4807–4811

Essays Biochem. (2009) **46**, 25–45; doi:10.1042/BSE0460003

3

S-Adenosylmethionine decarboxylase

Anthony E. Pegg[1]

Department of Cellular and Molecular Physiology, The Pennsylvania State University College of Medicine, PO Box 850, Hershey, PA 17033, U.S.A.

Abstract

S-Adenosylmethionine decarboxylase is a key enzyme for the synthesis of polyamines in mammals, plants and many other species that use aminopropyltransferases for this pathway. It catalyses the formation of *S*-adenosyl-1-(methylthio)-3-propylamine (decarboxylated *S*-adenosylmethionine), which is used as the aminopropyl donor. This is the sole function of decarboxylated *S*-adenosylmethionine. Its content is therefore kept very low and is regulated by variation in the activity of *S*-adenosylmethionine decarboxylase according to the need for polyamine synthesis. All *S*-adenosylmethionine decarboxylases have a covalently bound pyruvate prosthetic group, which is essential for the decarboxylation reaction, and have similar structures, although they differ with respect to activation by cations, primary sequence and subunit composition. The present chapter describes these features, the mechanisms for autocatalytic generation of the pyruvate from a proenzyme precursor and for the decarboxylation reaction, and the available inhibitors of this enzyme, which have uses as anticancer and anti-trypanosomal agents. The intricate mechanisms for regulation of mammalian *S*-adenosylmethionine decarboxylase activity and content are also described.

Introduction

Polyamines are small basic molecules found in virtually all living cells. Although diamines are frequently referred to as polyamines, they are really precursors

[1]*To whom correspondence should be addressed (email aep1@psu.edu).*

and polyamines are molecules with more than two positively charged nitrogen atoms. Two general pathways for synthesis of such polyamines have been described. By far the best known and most studied route requires the addition of aminopropyl groups catalysed by aminopropyltransferases that use dcAdoMet [decarboxylated AdoMet (*S*-adenosylmethionine); *S*-adenosyl-1-(methylthio)-3-propylamine] (Figure 1) as a substrate. This is the pathway used by many bacteria, archaea, fungi and higher organisms including mammals and plants [1,2], and was first characterized by Herbert and Celia Tabor using *Escherichia coli* [3]. The second pathway, which was described by Tait in 1976 [4], employs aspartate β-semialdehyde as the aminopropyl group donor in a fusion reaction to form carboxyspermidine or carboxynorspermidine, which is then decarboxylated. This pathway is present in *Vibrio cholerae* and many other bacteria [5].

The present chapter will describe in detail the properties and regulation of AdoMetDC (AdoMet decarboxylase), which is essential for the first pathway. The presence of a gene encoding AdoMetDC is therefore a diagnostic feature of organisms that use the aminopropyl transfer pathway. Supply of dcAdoMet limits the conversion of putrescine into the higher polyamines in such species and AdoMetDC is the key regulatory step in the biosynthetic pathway. Once converted into dcAdoMet, AdoMet is exclusively utilized for polyamine biosynthesis; there are no other known reactions that use dcAdoMet as a substrate, except for an acetylation reaction described below.

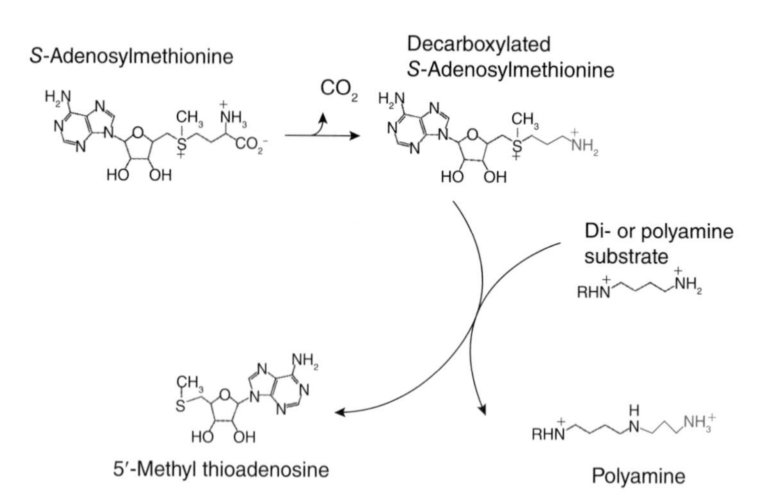

Figure 1. The AdoMetDC reaction and the synthesis of polyamines by aminopropyltransferases
The aminopropyl group is shown in blue. The aminopropyltransferase reaction shown is with putrescine (R=H) or spermidine R=$N^+H_3(CH_2)_3$ as a substrate. Known aminopropyltransferases can add aminopropyl groups to other substrates such as agmatine, 1,3-diaminopropane, 1,5-diaminopentane and a variety of polyamines made up of aminobutyl or aminopropyl units [2,9].

Content of dcAdoMet in cells

Since decarboxylation prevents the availability of AdoMet for methyl transfer reactions, the steady-state level of dcAdoMet is kept very low (approx. 1–2% of the AdoMet content in mammals) [6]. The content of dcAdoMet can be increased greatly in mammalian cells by reduction of polyamine synthesis. Treatment with DFMO (α-difluoromethylornithine), an inhibitor of ODC (ornithine decarboxylase) that shuts off production of putrescine, results in a huge increase in dcAdoMet leading to levels 3–4-fold greater than AdoMet itself [7,8]. This rise occurs because (a) there is no supply of the amine substrate for the aminopropyltransferase reactions, and (b) AdoMetDC is greatly induced in response to the fall in polyamines. A smaller, but still highly significant, rise in dcAdoMet (20–100-fold depending on the tissue) occurs in Gy mice, which have a deletion of the spermine synthase gene (D.E. McCloskey and A.E. Pegg, unpublished work).

In cases where dcAdoMet levels are greatly increased, there is also the appearance of an acetylated derivative [7,8]. This is not itself a substrate for the aminopropyltransferase reactions, but when polyamine synthesis is resumed, it rapidly disappears due to action of a deacetylase.

AdoMet can exist in two diastereoisomeric states with respect to its sulfonium ion. The *S* configuration, (*S*,*S*)-AdoMet, is the form that is produced enzymatically. However, under physiological conditions, it can spontaneously racemize at the sulfonium ion to the *R* form producing (*R*,*S*)-AdoMet. The (*S*,*S*)-AdoMet is the substrate for AdoMetDC and the reaction produces (*S*,*S*)-dcAdoMet, which is the isomer used by aminopropyltransferases [2,9]. Most methyltransferases also use (*S*,*S*)-AdoMet, but a unique homocysteine methyltransferase actually uses the (*R*,*S*)-AdoMet and thus prevents its accumulation [10]. (*S*,*S*)-dcAdoMet can also spontaneously racemize in aqueous solution at neutral pH [11], but it is not known what prevents accumulation of (*R*,*S*)-dcAdoMet *in vivo.* It is possible that most of the very low level of the nucleoside is bound by spermidine and spermine synthases, which together are present in almost equal concentrations to dcAdoMet, and the efficient utilization by these aminopropyltransferases is sufficient to prevent accumulation of the (*R*,*S*) form.

Strucure and function of AdoMetDC

Nature of the AdoMetDC reaction and activation by effectors of low molecular mass or binding of other proteins

All AdoMetDCs are synthesized as a proenzyme (π chain) and are made up of α and β subunits that are formed from it. The first AdoMetDC to be characterized was from *E. coli* [3]. This enzyme requires Mg^{2+} [12]. The second AdoMetDC was from mammals; it was found to be unaffected by Mg^{2+}, but required putrescine for maximal activity [13]. Subsequent work has found that putrescine-activated AdoMetDCs are widespread and present in fungi, protozoa and nematodes; however, AdoMetDCs in plants have no known effectors and are not activated by putrescine or by Mg^{2+} [14,15]. Quite recently, it has been found that AdoMetDCs in trypanosomes form a heterodimer with a catalytically inactive paralogue of the π chain called proenzyme (π′). This greatly stimulates the activity [16].

AdoMetDCs can be divided into five classes based on sequence at the cleavage site, structural considerations and their activation by cations (Table 1). Class I enzymes include those from bacterial and archaeal sources. All have αβ subunit structures, but may be monomers, dimers or tetramers of these units. They can be divided into Mg^{2+}-activated Class IA representing Gram-negative bacteria, and Class IB from Gram-positive bacteria and archaeabacteria, which have no requirement for Mg^{2+}. Class II enzymes include those from mammals, plants, fungi and trypanosomes. Those in Class IIA, such as AdoMetDCs from mammals, are $(\alpha\beta)_2$ dimers and are putrescine-activated. Class IIB enzymes, such as those from plants, are αβ monomers and are not activated by putrescine. Class IIC contains the heterodimeric enzymes from trypanosomes, which have the subunit structure αβπ′. The AdoMetDC from *Plasmodium falciparum* is a bifunctional protein with a C-terminal domain that encodes ODC [17].

Structural studies of AdoMetDCs

Crystal structures are now available for AdoMetDC proteins from three species: human, potato and *Thermotoga maritima*. Structures of the unprocessed proenzyme π chain [18] and the processed human enzyme in the presence of inhibitors and substrate analogues, as well as in the absence and presence of the putrescine activator, have been obtained [18–21]. Also, there are structures for the processed potato AdoMetDC [15] and the π proenzyme from *T. maritima* [22]. A model of the *P. falciparum* structure has been published [17].

Human AdoMetDC is an $(\alpha\beta)_2$ dimer from Class IIA (Figures 2A and 2B). Each αβ monomer is a novel four-layer α/β-sandwich fold, in which two central eight-stranded antiparallel β-sheets are flanked by several α-helices on each outer side. The β subunit contributes three β-strands, one α-helix and one short 3_{10}-helix. The α subunit contributes 13 β-strands, five α-helices and three short 3_{10}-helices. The six α-helices observed in the monomer are all amphipathic, packing tightly against the outer faces of the β sandwich. The region between the α-helices and the β-sheets is very hydrophobic. The overall structure of the potato enzyme is similar to human AdoMetDC, but the potato enzyme is an αβ monomer [15]. Residues at the dimer interface of human AdoMetDC are replaced in the potato structure and these substitutions create steric conflicts preventing dimerization.

There is very little overall primary sequence similarity between the Class I and the Class II AdoMetDCs. However, the crystal structure of the *T. maritima* AdoMetDC Class IB π proenzyme shows that the overall structure is remarkably similar to the Class II eukaryotic AdoMetDC αβ (Figures 2A and 2B), suggesting an evolutionary link between them [22]. The main differences between the human and *T. maritima* structures are a result of several insertions in the human protein, which occur primarily relative to the

Table 1. Classes of AdoMetDCs

The serine residue that is converted into pyruvate is shown in bold.

Species	π Chain size (kDa)	Subunits (number of amino acids) π	α	β	Activator	Cleavage site sequence	Oligomer structure
Class IA							
E. coli	30	264	152	111	Mg^{2+}	VVAHLDK**S**^{112}HICVH	$(\alpha\beta)_4$
C. acetobutylicum	32	274	155	118	Mg^{2+}	VVAHLDK**S**^{119}HICVH	Not reported
Class IB							
T. maritima	15	130	67	62	None	GVVVISE**S**^{63}HLTIH	$(\alpha\beta)_2$
M. jannaschii	14	124	60	63	None	GVAVLAE**S**^{64}HIAIH	$(\alpha\beta)_2$
B. subtilis	14	127	62	64	None	GVVVISE**S**^{65}HLTIH	Not reported
Class IIA							
Human	38	334	266	67	Putrescine	EAYVLSE**S**^{68}SMFVS	$(\alpha\beta)_2$
Yeast	46	396	308	87	Putrescine	DAFLLSE**S**^{88}SLFVF	$(\alpha\beta)_2$
Class IIB							
Potato	40	360	287	72	None	DSYVLSE**S**^{73}SLFVY	$\alpha\beta$
C. roseus	39	358	286	71	None	DSYVLSE**S**^{72}SLFVY	$\alpha\beta$
Class IIC							
T. brucei	44	369	283	85	π' Subunit	RSYVLSE**S**^{86}SLFVM	$\alpha\beta\pi'$
T. cruzi	42	370	284	85	π' Subunit	RSYVLTE**S**^{86}SLFVM	$\alpha\beta\pi'$

(A)

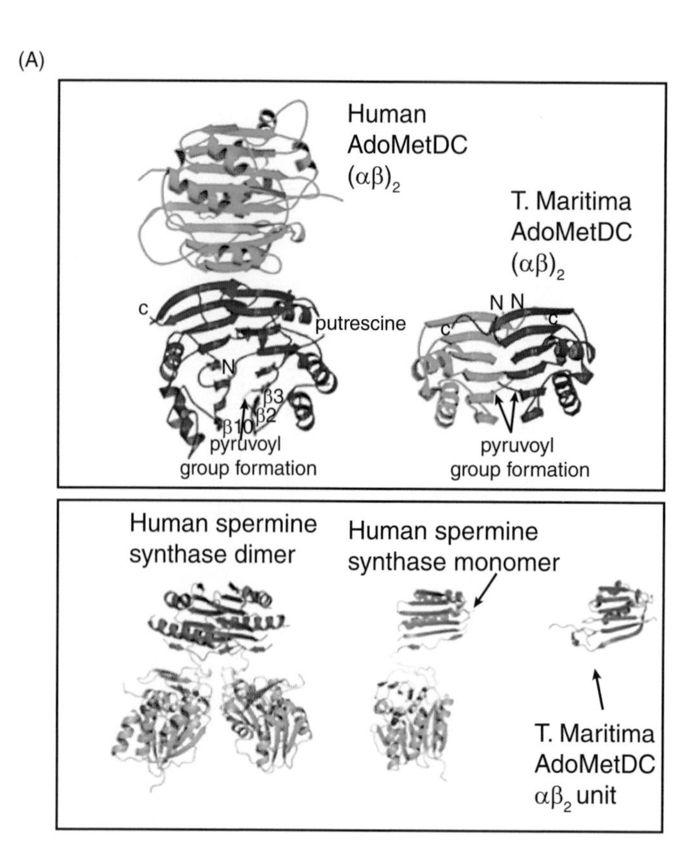
Human
AdoMetDC
$(\alpha\beta)_2$
T. Maritima
AdoMetDC
$(\alpha\beta)_2$
putrescine
pyruvoyl
group formation
pyruvoyl
group formation
Human spermine
synthase dimer
Human spermine
synthase monomer
T. Maritima
AdoMetDC
$\alpha\beta_2$ unit

(B)

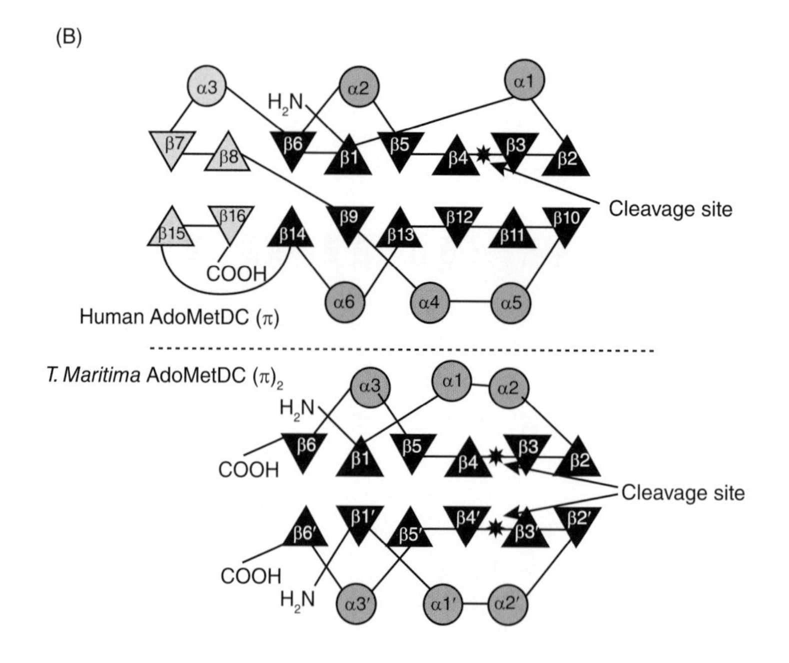
Cleavage site
COOH
Human AdoMetDC (π)
T. Maritima AdoMetDC $(\pi)_2$
COOH
Cleavage site
COOH

(C) Comparison of Human and *T. maritima* AdoMetDC sequences

```
                  β1           α1                       α2              β2
                ββββββββββββ   αααα                 ααααααααααααααα   ββββββββββ       ββ
T. maritima   -GRHLVAEFYECDREVL..........DNVQLIEQEMKQAAYESGATIVTSTFHRFLPYGVSGV (5-57)
N-H. sapiens  -EKLLEVWFSRQQ.....PDANQGSGDLR.TIPRSEWDILLKDVQCSIISVTKTD.....KQEA (8-62)
C-H. sapiens  -DQTLEILMSELDPAVMDQFYMKDGVTAKDVTRESGIRDLIP.....GSVIDATMFNPCGYSMN (174-231)

                β3      β4            β5              α3                β6
               βββ    βββββββββ     ββββββββββ     ααααααααααααα      ββββββββββββββ
T. maritima    VVISE.SHLTIHTWPE..YGYAAIDLFTCGEDVDPWKAFEHLKKALK....AKRV.HVVEHER- (58-112)
N-H. sapiens   YVLSE.SSMFVSK........RRFILKTCGT.TLLLKALVPLLKLARDYSGFDSIQSFFYSRK- (63-115)
C-H. sapiens   GMKSDGTYWTIHITPEPEFSYVSFETNLSQTS..YDDLIRKVVEVFK....PGKF.VTTLFVN- (232-287)
```

Comparison of *E. coli* and *T. maritima* AdoMetDC sequences

```
                   β1                            α1        α2               β2
               ββββββββββββ                      αααα ααααααααααααααα   βββββββββ    βββ
T. maritima    MKSLGRHLVAEFYECD..........REVLDNVQLIEQEMKQAAYESGATIVTSTFHRFLPYGVS (1-55)
E. coli       -TKSLSFCIYDICYAKTAEERDGYIAYIDELYNANRLTEILSETCSIIGANILNIARQDYEPQGAS (13-77)

                β3                                   β4                 β5
               βββββ                              ββββββββββ         ββββββββββ
T. maritima    GVVVISE............................SHLTIHTWPEYG........YAAIDLFTCG (56-84)
E. coli        VTILVSEEPVDPKLIDKTEHPGPLPETVVAHLDKSHICVHTYPESHPEGGLCTFRADIEVSTCG (78-141)

                    α3          β6
                ααααααααααα  ββββββββββββ
T. maritima    EDVDPWKAFEHLKKALKAKRVHVVEHERG-                              (85-113)
E. coli        V.ISPLKALNYLIHQLESDIVTIDYRVRG-                              (141-169)
```

Figure 2. Structures of AdoMetDC from human and *T. maritima* and of human spermine synthase

The upper part of (**A**) shows the human $(\alpha\beta)_2$ dimer of processed human AdoMetDC, which has one active site in each unit, and the structure of the *T. maritima* π chain which could give rise to an $(\alpha\beta)_2$ dimer with two active sites at the positions shown (see [22]). The lower part of (**A**) shows the dimer (native active form) and monomer of human spermine synthase [9], and the similarity of its N-terminal domain to AdoMetDC. (**B**) shows a comparative topology diagram of the arrangement of α helices (circles) and β sheets (triangles) of the human π and *T. maritima* π_2 chains. The unfilled segments (α3, β7, β8, β15 and β16) of the human protein are not present in the *T. maritima* structure. Note that the *T. maritima* structure is of the π_2 dimer and has two potential cleavage sites for pyruvate formation. The human sequence diverges in one of the two sites and forms only one pyruvate prosthetic group. (**C**) shows a sequence comparison between (i) portions of the human and *T. maritima* AdoMetDCs and (ii) portions of the *E. coli* and *T. maritima* AdoMetDCs. The alignment in (i) is based on the superimposition of the crystal structures and the α-helices and β-strands are marked. The alignment in (ii) is based on the sequence similarity between the *E. coli* and *T. maritima* AdoMetDCs, which can be used to generate a model of the *E. coli* structure for which experimental data is not available [22]. The conserved serine residue that forms the pyruvate ($hSer^{68}$, $tSer^{63}$ and $eSer^{112}$) is shown in red. The conserved histidine and serine residues playing key roles in the processing reactions are shown in blue ($hHis^{243}$, $tHis^{68'}$, $eHis^{117'}$ and $hSer^{229}$, $tSer^{55'}$ and $eSer^{77'}$). The conserved cysteine and glutamate residues playing key roles in catalysis and substrate binding are shown in green ($hCys^{82}$, $tCys^{83}$, $eCys^{140}$ and $hGlu^{247}$, $tGlu^{72'}$ and $eGlu^{121'}$). Also shown in green are the aromatic residues interacting with the adenine of the dcAdoMet substrate ($hPhe^{223}$ and the equivalent $tPhe^{49'}$ and $tTyr^{71'}$; $hPhe^{7}$ is not aligned here but the side-chains of residues $tTrp^{70'}$ and $eTyr^{119'}$ which are shown in green are in a similar position) in the active site. The rigid proline-rich loop present in Mg^{2+}-activated Class IA AdoMetDCs is shown with yellow shading. Figure adapted from [22], Toms, A.V., Kinsland, C., McCloskey, D.E., Pegg, A.E. and Ealick, S.E. (2004) Evolutionary links as revealed by the structure of *Thermotoga maritima* S-adenosylmethionine decarboxylase. J. Biol. Chem **279**, 33837–33846, JBC Online (http://www.jbc.org), with permission.

C-terminus of each *T. maritima* AdoMetDC monomer (see the topology diagram in Figure 2B). The insertions form two additional β-strands in both the N- and C-terminal halves of the protein, an additional α-helix in the N-terminal half, as well as a loop region that connects the N- and C-terminal halves of the human structure. Two of the additional β-strands are involved in the $(\alpha\beta)_2$ dimer interface of the human AdoMetDC. The dimer interface of the *T. maritima* AdoMetDC contains the active site, which is made up of residues from both $\alpha\beta$ units. There are likely to be two active sites in the $(\alpha\beta)_2$ dimer (Figures 2A and 2B) [22]. However, the only structure available is of the unprocessed π_2 chain and it is not certain that both sites do process, since processing of one site could influence that of the other. In light of recent data on the active form of the Trypanosoma AdoMetDCs being $\alpha\beta\pi$ [16], this question requires further study. A number of key residues involved in protein processing, substrate binding and catalysis are positioned in the structure in similar places and can readily be identified, not only in the *T. maritima* structure, but also in the Class IA *E. coli* sequence for which no structure is currently available (Figure 2C).

Remarkably, the recent determination of the structure of human spermine synthase shows that it has an N-terminal domain with a structure that is extremely similar to AdoMetDC, although there is little identity at the amino acid sequence level [9]. Spermine synthase is a homodimer and each monomer consists of a C-terminal domain, which contains the active site, a central domain made up of four β-strands and this N-terminal domain. The N-terminal domain is essential for activity and provides the majority of the dimer interface. It is not yet clear whether this is its sole function or whether there is any additional regulatory role of this domain. It has no AdoMetDC activity and lacks the serine residue needed for processing that is described in the next section. Similar fusion proteins of AdoMetDC-like and aminopropyltransferase sequences are found in vertebrates, Deuterostomia, some Protostomia and some bacteria, but not in nematodes, plants, fungi and all sequenced single-cell eukaryotes with the exception of the ciliate *Tetrahymena thermophila* [9]. The fusion protein is present in sea anemone, *Nematostella vectensis*, which is the last common ancestor of cnidarians and bilaterians [9].

Pyruvoyl prosthetic group and reaction mechanism

All AdoMetDCs currently characterized belong to a small group of decarboxylating enzymes that use a covalently bound pyruvate as a prosthetic group rather than the cofactor PLP (pyridoxal 5′- phosphate) typically employed in amino acid decarboxylation reactions [23]. Other examples of such pyruvoyl enzymes are phosphatidylserine decarboxylase, which forms phosphatidylethanolamine, aspartate-1-decarboxylase which forms β-alanine, and some, but not all, arginine and histidine decarboxylases.

The reactive carbonyl group of the pyruvate cofactor reacts with the substrate to form a Schiff base. This provides an electron sink via the amide carbonyl group of the pyruvate, and facilitates decarboxylation (Figure 3A). Protonation of the Cα of the product is then required to form the dcAdoMet, whose release regenerates the pyruvate. Structural and biochemical studies indicate that the side chain of residue Cys^{82} in the human enzyme (Figure 3B) is responsible for this reaction [20,24]. This amino acid is absolutely conserved in all AdoMetDCs (Figure 2C).

A disadvantage of the covalently bound pyruvate mechanism is that substrate-mediated transamination (see Figure 3A), which can occur as a minor side reaction via protonation of the incorrect Cα of the cofactor, leads to the release of an aldehyde product and the formation of alanine in place of

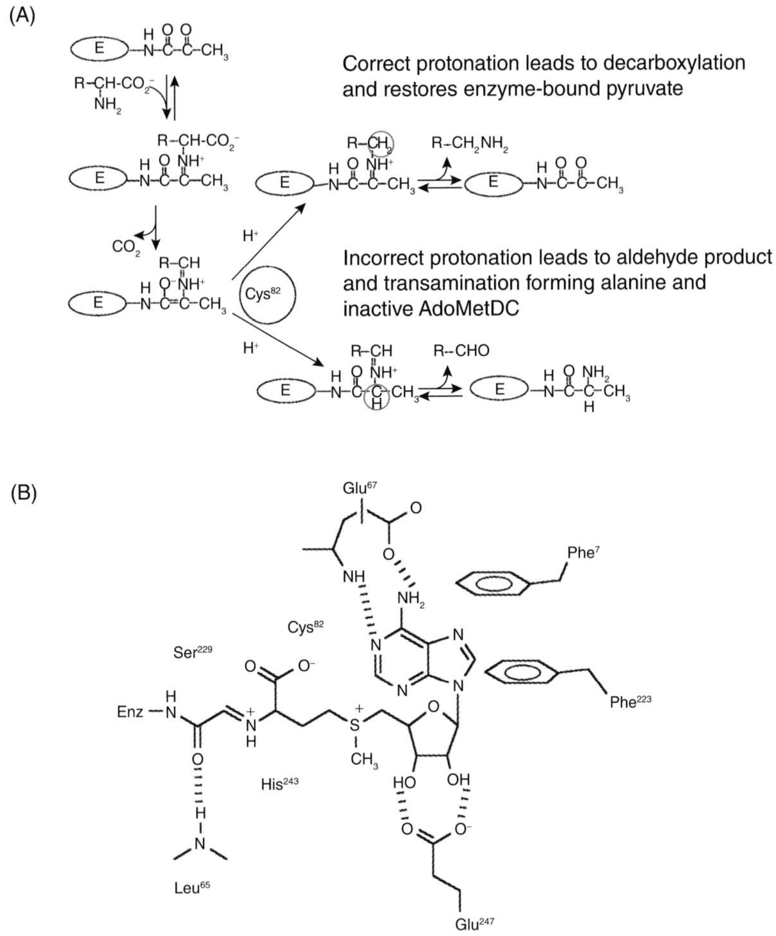

(A)

Correct protonation leads to decarboxylation and restores enzyme-bound pyruvate

Incorrect protonation leads to aldehyde product and transamination forming alanine and inactive AdoMetDC

(B)

Figure 3. Reaction of AdoMetDC
(**A**) shows the reaction brought about by AdoMetDC and all other decarboxylases with pyruvoyl prosthetic groups. Only the α carbon of the substrate amino acid and attached carboxyl and amino groups are shown (R represents the remainder of the AdoMet). A Schiff base is formed with the pyruvate attached to the enzyme (shown as E). After decarboxylation the intermediate is protonated (blue circle) using residue Cys^{82}. Correct protonation and product release (upper pathway) restores the pyruvate. Incorrect protonation (lower pathway) leads to an aldehyde product and transaminates the enzyme forming alanine. (**B**) shows a view of the AdoMet substrate Schiff base in the active site of human AdoMetDC based on the crystal structure [20].

the pyruvate. This irreversibly inactivates the enzyme since the resulting alanine residue is covalently attached. Such substrate-mediated transamination can also occur with enzymes using PLP as the cofactor, but the pyridoxamine formed can be replaced by a new PLP molecule restoring activity. When Cys^{82} is mutated to alanine in the human AdoMetDC, the reaction is greatly slowed and the rate of improper protonation and subsequent transamination is greatly increased [25]. This can be explained by another residue (probably His^{243}) that is not so well placed for correct protonation acting as the proton donor, either directly or via a water molecule.

Formation of the pyruvate

The mechanism by which the pyruvate is generated was first described by Snell and co-workers [26] more than 25 years ago on the basis of studies with histidine decarboxylase. The primary gene product is the single chain proenzyme (π chain). Formation of the active enzyme involves a self-maturation process of non-hydrolytic serinolysis in which the active-site pyruvoyl group is generated from an internal serine residue via an autocatalytic post-translational modification. The two subunits (α and β) are formed in this reaction (Figure 4A). The initial event in this pathway involves the formation of an internal ester bond using the side-chain -OH group of the serine residue. After cleavage of this bond, the oxygen atom from the serine -OH group is used to form the C-terminus of the β chain, while the remainder of the serine residue is converted into ammonia and the pyruvoyl group at the N-terminus of the α chain (Figure 4B). Snell and co-workers [26] used ^{18}O-labelling to provide convincing evidence for this mechanism. No ^{18}O was incorporated into the C-terminus of the β chain when cleavage was carried out in $H_2{}^{18}O$. When serine labelled with ^{18}O in the -OH group was used, the ^{18}O was transferred to the C-terminus of the β chain carboxylate.

The role of the protein in this autocatalytic mechanism is now understood in much more detail as a result of crystallographic studies of the protein structure and the results of site-directed mutagenesis of key residues. Much of this more recent work has been carried out with AdoMetDC. The human AdoMetDC π chain has 334 residues (38 kDa) and is converted into a 7.7 kDa (67 amino acids) β subunit and a 30.7 kDa α subunit, which contains 266 amino acids and the pyruvate group at its N-terminus (Figure 4A). The sequence VLSES^{68}SMFV surrounding the residue that forms the pyruvate (Ser^{68}) is highly conserved in AdoMetDC proenzymes from the Class II enzymes (Table 1).

As expected from the mechanism shown in Figure 4(B), replacement of the critical serine residue with any amino acid except cysteine or threonine completely prevents processing. Mutant AdoMetDCs with cysteine or threonine replacing this serine residue in the human or potato enzymes did process, albeit more slowly [14]. Replacement of the serine precursor with cysteine has no effect on the prosthetic group formed on cleavage, which is still pyruvate, but produces a thiocarboxylate group at the C-terminus of the β chain. This protein was still active [14]. Replacement of the serine with threonine generates an α-ketobutyrate prosthetic group. Although there is no reason that α-ketobutyrate cannot act in the same way as pyruvate and, indeed, the equivalent mutants in histidine decarboxylase or phosphatidylserine decarboxylase are active, the AdoMetDC mutant was not [14].

The initial rearrangement step to form the ester linkage requires a proton acceptor (Figure 4B). This is likely to be the side chain of Ser^{229} in human AdoMetDC. The structure of the π chain, which was determined using the non-processing S68A mutant, indicates that the -OH group is positioned

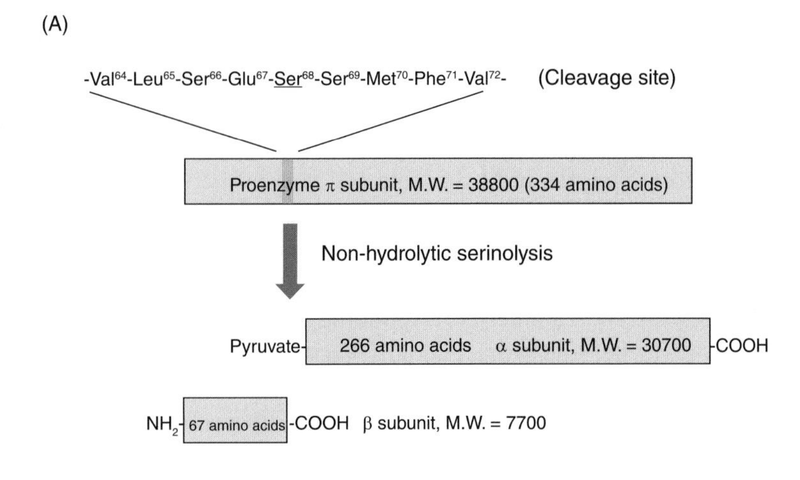

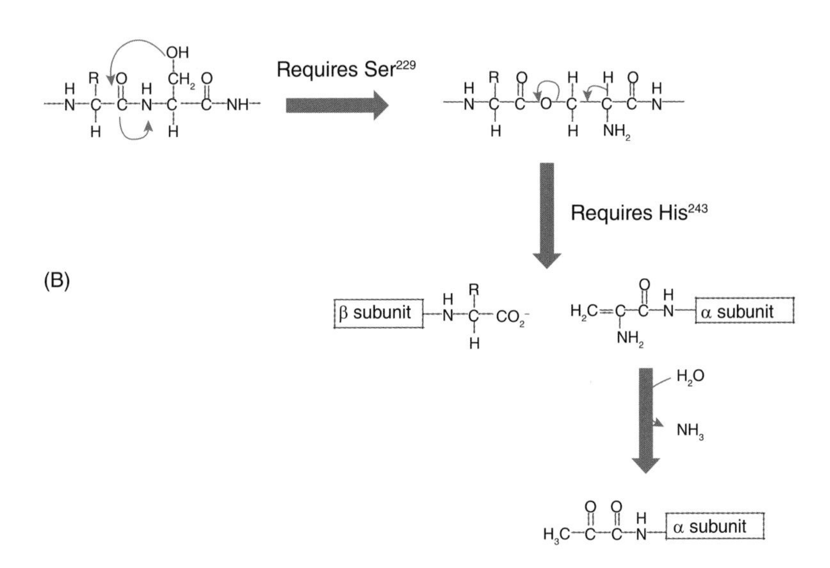

Figure 4. Formation of pyruvate

(**A**) shows the cleavage of the human AdoMetDC proenzyme to form the α and β subunits. (**B**) shows the mechanism by which cleavage occurs via non-hydrolytic serinolysis. M.W., molecular mass.

appropriately to form a hydrogen bond [18]. Mutant S229A failed to process, and mutant S229C cleaved very slowly, whereas mutant S229T processed normally [25]. Mutation of residue His[243] prevented normal processing, which stopped after the ester formation step. This was demonstrated by the finding that the H243A mutant remained as a π chain, but this chain could be cleaved into a modified α subunit with serine instead of pyruvate and the β subunit by mild treatment with hydroxylamine, which is known to break such ester bonds [25]. Determination of the crystal structure of the H243A mutant AdoMetDC protein has confirmed directly that it contains an ester bond between residues 67 and 68 [19]. A basic residue is needed to extract the hydrogen of the α-carbon of Ser^{68} in the ester to allow the β-elimination reaction that occurs to form the subunit with an N-terminal dehydroalanine residue (Figure 4B). His^{243} is therefore likely to fulfil this role and it is located in an appropriate position in the active site.

There is no significant similarity in the amino acid sequence surrounding the serine-cleavage site between the Class I and Class II AdoMetDCs (Table 1). However, conserved residues corresponding to His^{243} and Ser^{229} are positioned identically when the human and *T. maritima* sequences are compared (Figure 2C). Thus Ser$^{55'}$ is equivalent to Ser^{229} and His$^{68'}$ is equivalent to His^{243} [note that (′) means the residue comes from the other $\alpha\beta$ unit in the dimer shown in Figures 2A and 2B] [22]. Also, these residues can readily be identified in the *E. coli* AdoMetDC amino acid sequence, where they are provided by Ser77′ and His111′ (Figure 2C). A major difference between the Class IA and Class IB AdoMetDCs is the presence of a 27 amino acid insertion immediately prior to the cleavage site in the Class IA enzymes. This sequence, which contains five proline residues and a high concentration of negatively charged residues, would be expected to be very rigid and may play a role in the activation of the Class IA enzymes by Mg^{2+}.

Comparison of the structures of the processed wild-type human AdoMetDC, the unprocessed S68A mutant and the ester-containing H243A mutant have indicated that there is actually very little overall structural change [18,19]. Most of the changes occur at the Ser^{68} site of pyruvoyl group formation and in nearby residues Leu^{65}, Ser^{66}, Glu^{67}, Cys^{82}, Ser^{229} and His^{243}. The remainder of the enzyme, including the two antiparallel β-strands that are joined by residues 65–68, shows very little alteration in structure. The active-site pocket, which obviously includes Ser^{68}, must be compatible with the residues needed for catalysis and those needed for processing. This provides a serious constraint during the evolution of such pyruvoyl enzymes and may account for their relative rarity.

It is interesting that the first stage in the processing of pyruvoyl enzymes is similar to that involved in protein splicing, a post-translational editing process that removes an intein from a precursor protein and ligates the surrounding exteins to form the mature protein [27]. This process also begins with the side chain -OH or -SH of the conserved intein N-terminal serine or cysteine attacking the carbonyl carbon atom of the preceding amino acid, resulting in an ester or thioester bond at the N-terminal splice site. However, instead of undergoing the cleavage reaction, this is then attacked by the -OH/-SH group of the serine/threonine/cysteine from the second extein resulting in N-terminal cleavage and formation of a branched intermediate. This intermediate is resolved by cleavage of the peptide bond at the C-terminal splice site due to cyclization of the intein C-terminal asparagine residue and a spontaneous O–N or S–N acyl rearrangement establishes a peptide bond between the exteins.

The active-site pocket and substrate binding

Considerable insight into the catalytic reaction is provided by the crystal structures of human AdoMetDC with bound substrate analogues [20]. Both the α and β subunits contribute key residues to the active site. The AdoMet substrate is bound by interactions of the ribose -OH groups to Glu^{247} and the adenine ring is stacked between the aromatic rings of residues Phe^{7} and Phe^{223}. Remarkably, the adenosyl moiety of the bound substrate is in the *syn* conformation. In solution, AdoMet would be expected to adopt the *anti* structure and most of the known structures of AdoMet bound to proteins show the *anti* conformation. Glu^{247} is conserved in all AdoMetDCs; the two phenylalanine residues are sometimes replaced by other amino acids with aromatic ring side chains (see Figure 2C).

The critical importance of Glu^{247}, Phe^{7} and Phe^{223} has been confirmed by site-directed mutagenesis. Mutation of either phenylalanine residue to alanine results in a large increase in the K_m and a decrease in the k_{cat} [20]. Mutation of residue Glu^{247} to alanine increased the K_m for AdoMetDC by >10-fold and alteration to aspartate had a smaller 6-fold increase, whereas conversion into glutamine, which would still allow interaction with the ribose, had no significant effect. Residues Cys^{82}, Ser^{229} and His^{243} are positioned near the methionyl group of the substrate. As described above, Cys^{82} is the essential proton donor and its mutation to alanine has a profound effect on the reaction. The roles of Ser^{229} and His^{243} are less readily addressed by site-directed mutagenesis, since these residues are needed for processing to form the pyruvate, but their proximity to the substrate in the active site suggests that they do play a role in the catalytic reaction. The Ser^{229} mutations to threonine or cysteine, which do process, lead to a loss of >90% activity even from the processed protein [25].

The Cys^{82} residue in AdoMetDC, reacts readily with NO inactivating the enzyme. ODC is also inactivated by NO, which is therefore a potent inhibitor of polyamine synthesis [28]. Effects on polyamine synthesis may mediate some of the antiproliferative actions of NO.

Putrescine activation

The activation of AdoMetDC by putrescine links the supply of both substrates for spermidine synthase. Putrescine is therefore efficiently converted into spermidine. Mammalian cells and tissues contain much higher amounts of spermidine and spermine than putrescine, whereas in species that lack this putrescine activation, putrescine may be the predominant polyamine.

Mammalian AdoMetDC is activated by putrescine in two ways. The processing reaction is stimulated by putrescine [29], and the catalytic activity of the enzyme is enhanced [30–32]. Recent structural and biochemical studies indicate that there is one putrescine-binding site per αβ unit located a significant distance from the active site between two central β-sheets of the enzyme and near the dimer interface (Figure 2A). Putrescine binding to the $(\alpha\beta)_2$ dimer, which makes up the enzyme, is quite strongly co-operative [33]. The aliphatic chain of putrescine stacks against Phe^{111} and Phe^{285}. One end of putrescine is directly hydrogen-bonded to Asp^{174}, Glu^{15} and Thr^{176}. The other end is hydrogen-bonded through water molecules to Glu^{178}, Glu^{15}, Glu^{256} and Ser^{113}. These results are consistent with results of site-directed mutagenesis in which residues Asp^{174}, Glu^{178} or Glu^{256} were changed to the corresponding amides. Processing and activity of these mutants was not increased by putrescine [14,34].

The putrescine molecule is linked to the active site by several buried charged residues and its binding causes structural and electrostatic alterations at the active site. Binding of putrescine results in a reorganization of four aromatic residues (Phe^{285}, Phe^{315}, Tyr^{318} and Phe^{320}) and a conformational change in a loop formed by residues 312–320. This shields putrescine from the external solvent, enhancing its electrostatic and hydrogen-bonding effects [33].

The Class IIB AdoMetDC from potato is not activated by putrescine [14]. The overall structure of the potato protein is similar to that of the human, but the protein differs in two critical respects, which explains the lack of putrescine stimulation [15]. First, as described above, it is an $\alpha\beta$ monomer. Secondly, several amino acid substitutions of residues in the buried region where the putrescine-binding site is located in the human protein provide basic side chains that mimic the role of putrescine in the human enzyme. Thus the electrostatic interactions affecting the active site are produced by these residues without need for a ligand.

Inhibition of AdoMetDC

There is much interest in the polyamine biosynthetic pathway as a target for drugs that can be used as anti-protozoal, cancer chemotherapeutic and cancer chemoprevention agents. Several potent inhibitors of AdoMetDC have been designed and tested [21,35,36] (Figure 5). The first such inhibitor was MGBG [methylglyoxal bis(guanylhydrazone)], which was actually used as an anticancer drug in the 1960s before its ability to inhibit polyamine synthesis at the AdoMetDC step was discovered in 1972 [37]. Related diamidines, such as SAM486A (4-amidinoindan-1-one-2′-amidinohydrazone), which are more potent both as AdoMetDC inhibitors and as anticancer agents, were described in 1992 [37a]. The reason for the inhibition of AdoMetDC by these diamidines remained unclear until the crystal structures of AdoMetDC bound to the drugs were obtained [20]. These structures showed that the compounds are exactly the right length to span the entrance of the AdoMetDC active site via interactions of the two amidine groups at each end. One forms hydrogen bonds with Glu^{247} (which normally interacts with the ribose -OH groups of AdoMet) and the other bonds with the side chain of Ser^{229} and the main chain carbonyl group of Leu^{65}. The spacer groups of the inhibitors are located between the two phenylalanine residues at positions 7 and 223, which normally provide stacking interactions with the adenine ring of the AdoMet substrate (Figure 3B). SAM486A is able to interact more strongly than MGBG in this way by virtue of its aromatic rings, thus accounting for its greater potency.

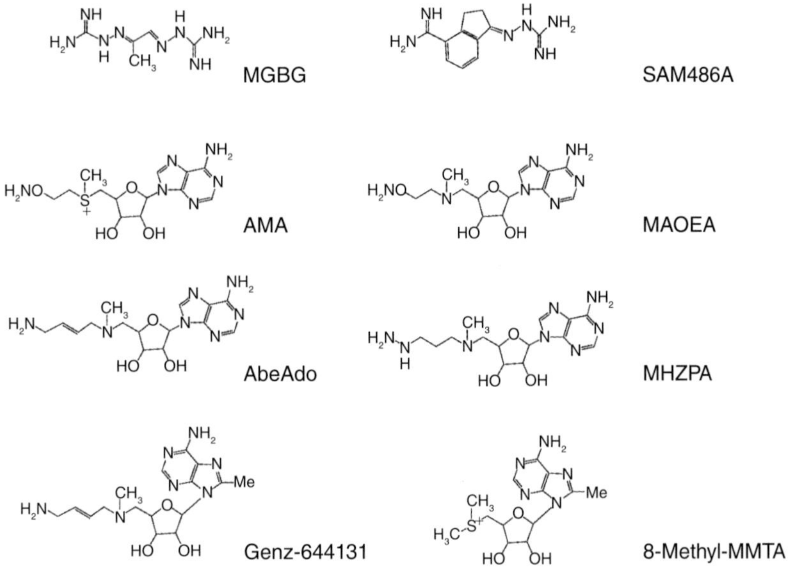

Figure 5. Structures of some important inhibitors of AdoMetDC
The compounds shown are: MGBG [methylglyoxal bis(guanylhydrazone)], SAM486A (previously described as CGP4864A; 4-amidinoindan-1-one-2′-amidinohydrazone), AMA [*S*-(5′-deoxy-5′-adenosyl)methylthioethylhydroxylamine], MAOEA {5′-deoxy-5′-[(2-aminooxyethyl)methylamino]adenosine}, AbeAdo (previously described as MDL73811; 5′-{[(Z)-4-amino-2-butenyl]methylamino}-5′-deoxyadenosine), MHZPA {5′-deoxy-5′-[(3-hydrazinopropyl)methylamino]adenosine}, Genz-644131 (5′-{[(Z)-4-amino-2-butenyl]methylamino}-5′-deoxy-8-methyladenosine) and 8-methyl-MMTA (5′-deoxy-5′-dimethylsulfonio-8-methyladenosine). The two 8-methyl derivatives (bottom row) are shown in the *syn* conformation.

A number of nucleoside derivatives related to AdoMet containing reactive aminooxy-, hydrazino- or hydrazido- groups such as MHZPA {5′-deoxy-5′-[(3-hydrazinopropyl)methylamino]adenosine}, MAOEA {5′-deoxy-5′-[(2-aminooxyethyl)methylamino]adenosine} and AMA [*S*-(5′-deoxy-5′-adenosyl) methylthioethylhydroxylamine] have also been shown to be irreversible inhibitors ([20,21,35] and references therein). They bind at the active site, forming a covalent linkage to the pyruvoyl prosthetic group and are potent inhibitors, but seem to be too reactive with other cellular compounds to be useful drugs. As with the substrate AdoMet, they adopt the *syn* conformation in the active site [20]. An 8-methyl substitution increases their efficiency by favouring this conformation [21]. Other AdoMet analogues, including some with such 8-methyl substitutions that cannot be decarboxylated, but bind reversibly to the active site, have also been described including 8-methyl-MMTA (5′-deoxy-5′-dimethylsulfonio-8-methyladenosine) [21,35]. These may be more useful as drugs despite being less potent.

The most promising irreversible inhibitor of AdoMetDC is probably Genz-644131 (5′-{[(Z)-4-amino-2-butenyl]methylamino}-5′-deoxy-8-methyla denosine), the recently described 8-methyl derivative of AbeAdo (see below)

[36]. Genz-644131 has good potency, biostability and brain penetration and cured trypanosome infections in mice.

AbeAdo (5′-{[(Z)-4-amino-2-butenyl]methylamino}-5′-deoxyadenosine), also known as MDL73811, is a stable molecule that was designed 20 years ago as an enzyme-activated irreversible inhibitor of AdoMetDC [38]. The postulated mechanism of action of AbeAdo involves the formation of a Schiff base between the drug and the pyruvate, followed by an enzyme-mediated abstraction of a proton forming a conjugated imine, which could then react with a nucleophilic residue of AdoMetDC. AbeAdo is indeed a potent inactivator of AdoMetDC, but when recombinant human AdoMetDC was allowed to react with AbeAdo, the enzyme was inactivated by a transamination of the pyruvate group [39], suggesting that its primary mechanism of action was actually to increase the incorrect protonation shown in Figure 3(A). Irrespective of the mechanism of action, the increased potency of Genz-644131 over the parent compound is explained by the greater likelihood of the compound being in the *syn* configuration.

Regulation of AdoMetDC content

In addition to the activation of AdoMetDC proenzyme processing and enzyme activity by putrescine described above, the amount of AdoMetDC protein is highly regulated to respond to the cellular need for polyamines. Therefore many physiological processes leading to increased growth raise the AdoMetDC content. Regulation can occur at multiple steps including transcription of the *AdoMetDC* gene, mRNA translation and protein turnover (Figure 6) [31,32]. The AdoMetDC amount is negatively regulated by increased polyamine content at all of these steps. Therefore drugs or pathophysiology that reduce polyamine levels increase the AdoMetDC content.

The mechanism by which transcription is increased is not well understood, although the gene contains a number of binding sites for key cell-regulatory transcription factors and may contain a polyamine-responsive element [32]. The best understood regulatory step in the synthesis of AdoMetDC is the translational regulation [40,41]. This regulation is achieved by virtue of small ORFs (open reading frames) in the 5′-UTR (untranslated region) of the mRNA. Mammalian AdoMetDC mRNAs contain such an ORF, which is highly conserved and encodes the hexapeptide MAGDIS [40]. This ORF is located only 14 nucleotides from the 5′ end of the mRNA. Ribosomes binding to it stall just prior to termination. The stalling is actually related to the encoded MAGDIS sequence. Unless this is prevented, access to the downstream ORF that encodes the enzyme is blocked. Factors, such as polyamine levels affecting AdoMetDC synthesis, alter the efficiency of this process and modulate the translation of the downstream AdoMetDC reading frame in this way [40]. Plant AdoMetDC mRNAs actually contain two conserved overlapping upstream ORFs in their 5′ leader sequence encoding peptides of two or three and approx. 50 amino acids respectively. The ORF encoding the smaller peptide, which is located 5′ to the larger ORF, is not inhibitory, but occludes

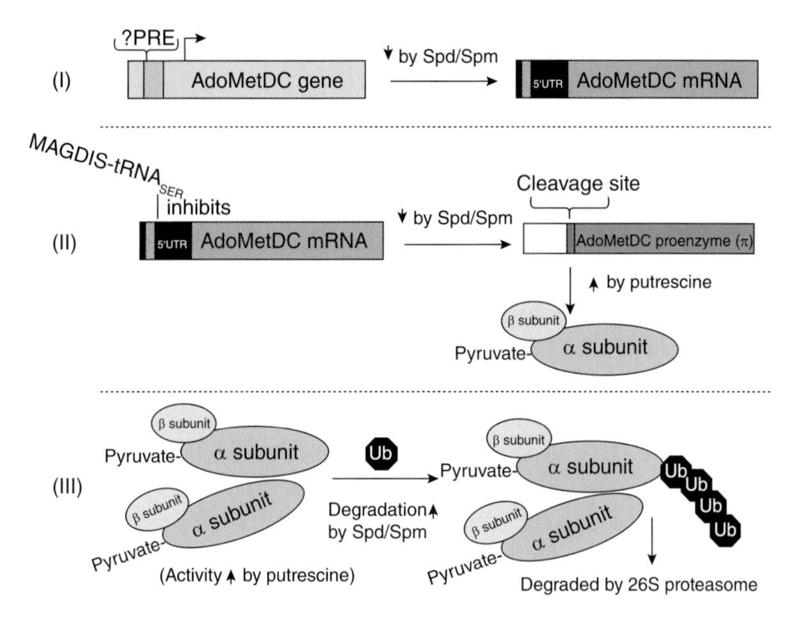

Figure 6. Regulation of mammalian AdoMetDC
(**I**) Transcription of the AdoMetDC gene is negatively regulated by spermidine and spermine (Spd/Spm). This may involve a polyamine-responsive element (PRE) in the gene. (**II**) Synthesis of AdoMetDC protein involves the 5′-UTR of the mRNA, where a small upstream ORF encoding the peptide MAGDIS is located. Synthesis of the proenzyme (π subunit) encoded by the main ORF is negatively regulated by Spd/Spm. The π subunit undergoes a spontaneous processing reaction to form the α and β subunits, which is stimulated by putrescine. (**III**) Activity of the final $(\alpha\beta)_2$ enzyme is increased by putrescine. Degradation of the enzyme by the proteasome requires polyubiquitination and is increased when Spd/Spm levels are high. The site of ubiquitin (Ub) attachment is not known. It is possible that transamination precedes ubiquitination.

the initiation codon of the second ORF. At high polyamine concentrations, this tiny ORF is ignored and binding of ribosomes to the second ORF translates it, giving sequence-dependent translational repression as for the MAGDIS sequence [41]. Upsteam ORFs that presumably have similar regulatory properties are found in the *AdoMetDC* mRNAs from many species [42].

The mammalian AdoMetDC protein turns over rapidly via degradation by the 26S proteasome after polyubiquitination [32,43]. This turnover is more rapid when polyamine levels are high. It is possible that the degradation of the inactive transaminated form of the enzyme produced by incorrect catalytic turnover, as described in Figure 3(A), is related to this. Increased flux through the AdoMetDC enzyme, which is needed for a high rate of polyamine synthesis, would increase the extent of such substrate-mediated transamination. Evidence that the polyubiquitination of the AdoMetDC protein and proteasomal degradation are enhanced after transamination has been published [44]. The reversible inhibitors described above that block the active site, such as MGBG and SAM486A, greatly stabilize AdoMetDC and thus have the unwanted effect of increasing its concentration, whereas treatment with AbeAdo causes a loss of the protein [32,44]. These observations are compatible with the model in which transamination precedes degradation. The site of ubiquitination is currently unknown, but it may be speculated that the formation of alanine at the N-terminus of the α subunit generates such a site or causes a structural alteration allowing the recognition by ubiquitin ligases.

Conclusions

AdoMetDC is a potentially important target for drug design. The proenzyme processing leading to the formation of its pyruvate prosthetic group and its catalytic mechanism are relatively well understood. Some further work on the structures of the Class IA and IIC AdoMetDC enzymes for which only models are currently available would be helpful, but current knowledge based on the Class IIA, IIB and IB enzymes combined with the database provided by the inhibitors currently available should allow for the design and synthesis of more potent and species specific inactivators. It is surprising that little attention has yet been given to the opportunities for drug design provided by the processing reaction; blocking this process or preferably causing an abortive cleavage without pyruvate formation is an attractive and novel route to interfering with AdoMetDC activity. The intricate regulation of AdoMetDC content and activity and its importance are an ongoing research area that is vitally important to a complete understanding of the control of polyamine levels. Clearly, more needs to be determined on the transcriptional regulation and on the mechanism for the rapid turnover of the AdoMetDC protein. The potential interaction of AdoMetDC with spermine synthase, and the extent to which the proenzyme (π′) subunit discovered in trypanosomes regulates activity in other species are exciting new areas for further study.

Summary

- *AdoMetDC is a key enzyme for the synthesis of polyamines in mammals, plants and many other species that use aminopropyltransferases for this synthesis.*
- *Since decarboxylation prevents the availability of AdoMet for methyl transfer reactions, the steady-state level of its decarboxylated derivative is kept very low.*
- *Some AdoMetDCs, such as those from mammals, are activated by putrescine, providing a means by which putrescine is efficiently converted into spermidine.*
- *AdoMetDC is very highly regulated in response to the need for polyamine production; regulation occurs at the levels of transcription, translation, proenzyme processing, enzyme activity and degradation.*
- *AdoMetDCs use a covalently bound pyruvate as a prosthetic group.*
- *The pyruvate prosthetic group is formed from an internal serine residue in a spontaneous processing reaction, which converts a proenzyme into the α and β subunits of the enzyme.*

- *Substrate AdoMet is bound at the active site in the usually unfavoured syn conformation.*
- *Despite very little sequence identity at the amino acid levels, the crystal structures of AdoMetDCs from different species are very similar.*
- *Inhibitors of AdoMetDC have been synthesized and shown to be potentially useful anticancer and anti-trypanosomal agents.*

Acknowledgements

This chapter is dedicated to the memory of Guy Williams-Ashman who persuaded me to work on polyamine biosynthesis and AdoMetDC more than 40 years ago. I thank all former and current colleagues laboratory members who have worked on AdoMetDC, and Dr Natalia A. Loktionova for help with the present chapter. I apologize that many important references have been omitted owing to limitations on the number of citations.

Funding

The work in the author's laboratory on polyamines is supported by the National Institutes of Health [grant numbers CA-018138, GM-26290]

References

1. Wallace, H.M., Fraser, A.V. and Hughes, A. (2003) A perspective of polyamine metabolism. Biochem. J. **376**, 1–14
2. Ikeguchi, Y., Bewley, M. and Pegg, A.E. (2006) Aminopropyltransferases: function, structure and genetics. J. Biochem. **139**, 1–9
3. Tabor, H., Rosenthal, S.M. and Tabor, C.W. (1958) The biosynthesis of spermidine and spermine from putrescine and methionine. J. Biol. Chem. **233**, 907–914
4. Tait, G.H. (1976) A new pathway for the biosynthesis of spermidine. Biochem. Soc. Trans. **4**, 610–612
5. Lee, J., Sperandio, V., Frantz, D.E., Longgood, J., Camilli, A., Phillips, M.A. and Michael, A.J. (2009) An alternative polyamine biosynthetic pathway is widespread in bacteria and essential for biofilm formation in *Vibrio cholerae*. J. Biol. Chem. **284**, 9899–9907
6. Hibasami, H., Hoffman, J.L. and Pegg, A.E. (1980) Decarboxylated *S*-adenosylmethionine in mammalian cells. J. Biol. Chem. **255**, 6675–6678
7. Pegg, A.E., Wechter, R.S., Clark, R.S., Wiest, L. and Erwin, B.G. (1986) Acetylation of decarboxylated *S*-adenosylmethionine by mammalian cells. Biochemistry **25**, 379–384
8. Wagner, J., Hirth, Y., Piriou, F., Zakett, D., Claverie, N. and Danzin, C. (1985) N-Acetyl decarboxylated *S*-adenosylmethionine, a new metabolite of decarboxylated *S*-adenosylmethionine: isolation and characterization. Biochem. Biophys. Res. Commun. **133**, 546–553
9. Wu, H., Min, J., Zeng, H., McCloskey, D.E., Ikeguchi, Y., Loppnau, P., Michael, A.J., Pegg, A.E. and Plotnikov, A.N. (2008) Crystal structure of human spermine synthase: implications of substrate binding and catalytic mechanism. J. Biol. Chem. **283**, 16135–16146
10. Vinci, C.R. and Clarke, S.G. (2007) Recognition of age-damaged (*R,S*)-adenosyl-L-methionine by two methyltransferases in the yeast *Saccharomyces cerevisiae*. J. Biol. Chem. **282**, 8604–8612
11. Dejima, H., Kobayashi, M., Takasaki, H., Takeda, N., Shirahata, A. and Samejima, K. (2003) Synthetic decarboxylated *S*-adenosyl-L-methionine as a substrate for aminopropyl transferases. Biol. Pharm. Bull. **26**, 1005–1008

12. Lu, Z.J. and Markham, G.D. (2007) Metal ion activation of S-adenosylmethionine decarboxylase reflects cation charge density. Biochemistry **46**, 8172–8180
13. Pegg, A.E. and Williams-Ashman, H.G. (1968) Stimulation of the decarboxylation of S-adenosylmethionine by putrescine in mammalian tissues. Biochem. Biophys. Res. Commun. **30**, 76–82
14. Xiong, H., Stanley, B.A., Tekwani, B.L. and Pegg, A.E. (1997) Processing of mammalian and plant S-adenosylmethionine decarboxylase proenzymes. J. Biol. Chem. **272**, 28342–28348
15. Bennett, E.M., Ekstrom, J.L., Pegg, A.E. and Ealick, S.E. (2002) Monomeric S-adenosylmethionine decarboxylase from potato provides an alternative to putrescine stimulation. Biochemistry **41**, 14509–14517
16. Willert, E.K., Fitzpatrick, R. and Phillips, M.A. (2007) Allosteric regulation of an essential trypanosome polyamine biosynthetic enzyme by a catalytically dead homolog. Proc. Natl. Acad. Sci. U.S.A. **104**, 8275–8280
17. Wells, G.A., Birkholtz, L.M., Joubert, F., Walter, R.D. and Louw, A.I. (2005) Novel properties of malarial S-adenosylmethionine decarboxylase as revealed by structural modelling. J. Mol. Graph. Model. **24**, 307–318
18. Tolbert, W.D., Zhang, Y., Bennett, E.M., Cotter, S.E., Ekstrom, J.L., Pegg, A.E. and Ealick, S.E. (2003) Mechanism of human S-adenosylmethionine decarboxylase proenzyme processing as revealed by the structure of the S68A mutant. Biochemistry **42**, 2386–2395
19. Ekstrom, J.L., Tolbert, W.D., Xiong, H., Pegg, A.E. and Ealick, S.E. (2001) Structure of a human S-adenosylmethionine decarboxylase self-processing ester intermediate and mechanism of putrescine stimulation of processing as revealed by the H243A mutant. Biochemistry **40**, 9495–9504
20. Tolbert, W.D., Ekstrom, J.L., Mathews, I.I., Secrist, J.A. I., Kapoor, P., Pegg, A.E. and Ealick, S.E. (2001) The structural basis for substrate specificity and inhibition of human S-adenosylmethionine decarboxylase. Biochemistry **40**, 9484–9494
21. McCloskey, D.E., Bale, S., Secrist, J.A., Tiwari, A., Moss, T.H., Valiyaveettil, J., Brooks, W.H., Guida, W.C., Pegg, A.E. and Ealick, S.E. (2009) New insights into the design of inhibitors of human S-adenosylmethionine decarboxylase: studies of adenine C_8 substitution in structural analogues of S-adenosylmethionine. J. Med. Chem. **52**, 1388–1407
22. Toms, A.V., Kinsland, C., McCloskey, D.E., Pegg, A.E. and Ealick, S.E. (2004) Evolutionary links as revealed by the structure of *Thermotoga maritima* S-adenosylmethionine decarboxylase. J. Biol. Chem. **279**, 33837–33846
23. Hackert, M.L. and Pegg, A.E. (1997) Pyruvoyl-dependent enzymes. In Comprehensive Biological Catalysis (Sinnott, M.L., ed.), pp. 201–216, Academic Press, London
24. Xiong, H., Stanley, B.A. and Pegg, A.E. (1999) Role of cysteine-82 in the catalytic mechanism of human S-adenosylmethionine decarboxylase. Biochemistry **38**, 2462–2470
25. Xiong, H. and Pegg, A.E. (1999) Mechanistic studies of the processing of human S-adenosylmethionine decarboxylase proenzyme. Isolation of an ester intermediate. J. Biol. Chem. **274**, 35059–35066
26. Recsei, P.A., Huynh, Q.K. and Snell, E.E. (1983) Conversion of prohistidine decarboxylase to histidine decarboxylase: peptide chain cleavage by nonhydrolytic serinolysis. Proc. Natl. Acad. Sci. U.S.A. **80**, 973–977
27. Paulus, H. (2002) Protein splicing and related forms of protein autoprocessing. Annu. Rev. Biochem. **69**, 447–496
28. Hillary, R.A. and Pegg, A.E. (2003) Decarboxylases involved in polyamine biosynthesis and their inactivation by nitric oxide. Biochim. Biophys. Acta **1647**, 161–166
29. Kameji, T. and Pegg, A.E. (1987) Effect of putrescine on the synthesis of S-adenosylmethionine decarboxylase. Biochem. J. **243**, 285–288
30. Pegg, A.E. and Williams-Ashman, H.G. (1969) On the role of S-adenosyl-L-methionine in the biosynthesis of spermidine by the rat prostate. J. Biol. Chem. **244**, 682–693

31. Stanley, B.A. (1995) Mammalian S-adenosylmethionine decarboxylase regulation and processing. In Polyamines: Regulation and Molecular Interaction (Casero, Jr, R.A., ed.), pp. 27–75, R.G. Landes Company, Austin
32. Pegg, A.E., Xiong, H., Feith, D. and Shantz, L.M. (1998) S-Adenosylmethionine decarboxylase: structure, function and regulation by polyamines. Biochem. Soc. Trans. **26**, 580–586
33. Bale, S., Lopez, M.M., Makhatadze, G.I., Fang, Q., Pegg, A.E. and Ealick, S.E. (2008) Structural basis for putrescine activation of human S-adenosylmethionine decarboxylase. Biochemistry **47**, 13404–13417
34. Stanley, B.A., Shantz, L.M. and Pegg, A.E. (1994) Expression of mammalian S-adenosylmethionine decarboxylase in *E. coli*. Determination of sites for putrescine activation of activity and processing. J. Biol. Chem. **269**, 7901–7907
35. Pegg, A.E. and McCann, P.P. (1992) S-Adenosylmethionine decarboxylase as an enzyme target for therapy. Pharmacol. Ther. **56**, 359–377
36. Barker, Jr, R.H., Liu, H., Hirth, B., Celatka, C.A., Fitzpatrick, R., Xiang, Y., Willert, E.K., Phillips, M.A., Kaiser, M., Bacchi, C.J. et al. (2009) Novel S-adenosylmethionine decarboxylase inhibitors for the treatment of human African trypanosomiasis. Antimicrob. Agents Chemother. **53**, 2052–2058
37. Williams-Ashman, H.G. and Schenone, A. (1972) Methylglyoxal bis(guanylhydrazone) as a potent inhibitor of mammalian and yeast S-adenosylmethionine decarboxylases. Biochem. Biophys. Res. Commun. **46**, 288–295
37a. Regenass, U., Caravatti, G., Mett, H., Stanek, J., Schneider, P., Müller, M., Matter, A., Vertino, P. and Porter, C.W. (1992) New S-adenosylmethionine decarboxylase inhibitors with potent antitumor activity. Cancer Res. **52**, 4712--4718
38. Casara, P., Marchal, P., Wagner, J. and Danzin, C. (1989) 5′-{[(Z)-4-Amino-2-butenyl]methylamino}-5′-deoxyadenosine: a potent enzyme-activated irreversible inhibitor of S-adenosyl-L-methionine decarboxylase from *Escherichia coli*. J. Am. Chem. Soc. **111**, 9111–9113
39. Shantz, L.M., Stanley, B.A., Secrist, J.A. and Pegg, A.E. (1992) Purification of human S-adenosylmethionine decarboxylase expressed in *Escherichia coli* and use of this protein to investigate the mechanism of inhibition by the irreversible inhibitors, 5′-deoxy-5′-[(3-hydrazino propyl)methylamino]adenosine and 5′{[(Z)-4-amino-2-butenyl]methylamino-5′-deoxyadenosine. Biochemistry **31**, 6848–6855
40. Raney, A., Law, G.L., Mize, G.J. and Morris, D.R. (2002) Regulated translation termination at the upstream open reading frame in S-adenosylmethionine decarboxylase mRNA. J. Biol. Chem. **277**, 5988–5994
41. Hanfrey, C., Elliott, K.A., Franceschetti, M., Mayer, M. J., Illingworth, C. and Michael, A.J. (2005) A dual upstream open reading frame-based autoregulatory circuit controlling polyamine-responsive translation. J. Biol. Chem. **280**, 39229–39237
42. Morris, D.R. and Geballe, A.P. (2001) Upstream open reading frames as regulators of mRNA translation. Mol. Cell. Biol. **20**, 8635–8642
43. Kahana, C. (2007) Ubiquitin dependent and independent protein degradation in the regulation of cellular polyamines. Amino Acids **33**, 225–230
44. Yerlikaya, A. and Stanley, B.A. (2004) S-Adenosylmethionine decarboxylase degradation by the 26S proteasome is accelerated by substrate-mediated transamination. J. Biol. Chem. **279**, 12469–12478

Essays Biochem. (2009) **46**, 47–61; doi:10.1042/BSE0460004

4

Regulation of cellular polyamine levels and cellular proliferation by antizyme and antizyme inhibitor

Chaim Kahana[1]

Department of Molecular Genetics, The Weizmann Institute of Science, Rehovot 76100, Israel

Abstract

Polyamines are small aliphatic polycations present in all living cells. Polyamines are essential for cellular viability and are involved in regulating fundamental cellular processes, most notably cellular growth and proliferation. Being such central regulators of fundamental cellular functions, the intracellular polyamine concentration is tightly regulated at the levels of synthesis, uptake, excretion and catabolism. ODC (ornithine decarboxylase) is the first key enzyme in the polyamine biosynthesis pathway. ODC is characterized by an extremely rapid intracellular turnover rate, a trait that is central to the regulation of cellular polyamine homoeostasis. The degradation rate of ODC is regulated by its end-products, the polyamines, via a unique autoregulatory circuit. At the centre of this circuit is a small protein called Az (antizyme), whose synthesis is stimulated by polyamines. Az inactivates ODC and targets it to ubiquitin-independent degradation by the 26S proteasome. In addition, Az inhibits uptake of polyamines. Az itself is regulated by another ODC-related protein termed AzI (antizyme inhibitor). AzI is highly homologous with ODC, but it lacks ornithine-decarboxylating activity. Its ability to serve as a

[1]*To whom correspondence should be addressed (chaim.kahana@weizmann.ac.il).*

regulator is based on its high affinity to Az, which is greater than the affinity Az has to ODC. As a result, it interferes with the binding of Az to ODC, thus rescuing ODC from degradation and permitting uptake of polyamines.

Introduction

The polyamines spermidine, spermine, and their diamine precursor, putrescine, are ubiquitous physiological cations. The polyamines fulfil a crucial role in regulating various fundamental cellular processes, most notably cell growth and proliferation. Depletion of cellular polyamines results in cessation of cellular proliferation, which can be completely reverted upon re-addition of exogenous polyamines. Although their explicit mechanism of action is still unknown, it is likely that their evenly distributed positive charges, which are capable of interacting with negative charges present in various cellular macromolecules, plays a key role in their ability to regulate cellular processes. Interactions with polyamines increase the melting temperature and condensation state of DNA. Polyamines can convert B-DNA into Z-DNA and A-DNA and affect binding of some transcription factors to DNA. Polyamine depletion inhibits both DNA and protein synthesis. The effect on protein synthesis may be mediated, at least in part, by an unusual covalent modification, termed hypusination, of the putative translation initiation factor eIF-5A (eukaryotic initiation factor 5A) [1]. Polyamines have also been demonstrated to act as activators and inhibitors of various cation channels [2–5] and have been suggested to modulate synthesis of nitric oxide [6]. However, the most notable feature of polyamines is their involvement in regulating proliferation of normal and malignant cells [7]. Depletion of cellular polyamine levels affects the expression of a number of growth-regulated/regulating genes. However, it is unknown whether these are the only changes in gene expression caused by changes in polyamine levels, and which of these changes affect growth, and which are actually an outcome of the growth arrest. Cell viability is affected both when polyamines are limiting and when they are present at excessive levels. Therefore their intracellular concentration must be maintained within a narrow optimal concentration. This is achieved by maintaining a balance between synthesis, catabolism, uptake and excretion. ODC (ornithine decarboxylase) is a highly regulated key enzyme in the polyamine biosynthesis pathway. ODC is subject to regulation by its end-products, the polyamines, through an autoregulatory circuit. At the centre of this circuit is a polyamine-induced protein termed Az (antizyme), which is itself regulated by an inactive ODC-related protein termed AzI (antizyme inhibitor).

ODC

ODC is a PLP (pyridoxal phosphate)-dependent enzyme that decarboxylates ornithine to form putrescine. In its active form, ODC exists as a homodimer with a two-fold symmetry. The monomer retains no enzymatic activity. The homodimer contains two symmetrical active sites located at the interface between the two subunits, with residues from each subunit contributing to the formation of each active site (for a review, see [8]). The association between the two subunits is rather weak, leading to a high rate of association and dissociation, a situation that, as will be described below, is central for its regulation.

ODC is a highly regulated enzyme that responds rapidly and dramatically to a variety of growth-promoting stimuli. The increase in ODC activity is correlated with an increase in the level of the enzyme protein. The increase in the amount of ODC is regulated at the levels of transcription, transcript stability and translatability. A key element enabling regulation by these regulatory mechanisms is the rapid turnover rate of ODC. In fact, ODC is one of the most rapidly degraded proteins in eukaryotic cells. Interestingly, ODC is not degraded at a constant rate, but rather its degradation rate is regulated by the intracellular concentration of polyamines. As mentioned above, two proteins regulate the degradation rate of ODC. The first, Az, whose synthesis is stimulated by polyamines, targets ODC to ubiquitin-independent degradation by the 26S proteasome. The second, AzI, negates the ability of Az to promote ODC degradation. Interestingly, Az and AzI also affect the process of polyamine uptake through the plasma membrane. However, neither the exact mechanism of polyamine uptake, nor the way it is affected by Az, are presently known.

Az

Az is a small protein originally described as a polyamine-induced ODC inhibitory activity. It is positioned at the centre of an autoregulatory circuit that regulates cellular polyamine levels.

Az synthesis

Az is encoded by two independent ORFs (open reading frames; ORF1 and ORF2) [9,10]. Translation that starts at one of two possible in-frame initiation codons is terminated shortly thereafter at an in-frame stop codon. In order to produce a functional full-length Az, the ribosomes that scan ORF1 must be subverted to the +1 reading frame (ORF2) (Figure 1). Comparison of the sequence of the mature functional Az protein with the sequence of the encoding mRNA reveals that the actual frameshifting event occurs while the scanning ribosome encounters the last codon of ORF1. In most cases, the ribosome slips one nucleotide forward, encoding the first amino acid (aspartic acid, GAU) of ORF2. Polyamines stimulate the efficiency of this +1 frameshifting event. The reliance of frameshifting efficiency on the polyamine concentration serves as an intracellular sensing mechanism for the free intracellular polyamine pool.

The most important segment of *Az* mRNA that is required for frameshifting is the sequence at which frameshifting occurs. This segment contains two elements, the stop codon that halts the scanning ribosome, and the sequence

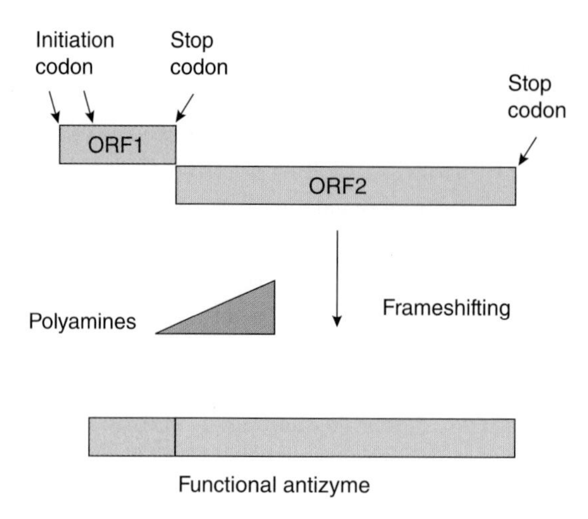

Figure 1. Synthesis of Az is regulated by a polyamine-stimulated ribosomal frameshifting
Translation of Az mRNA is initiated at one of two in-frame initiation codons. Shortly following translation initiation the scanning ribosome encounters an in-frame stop codon, generating a short non-functional polypeptide. To obtain a functional full-length Az the ribosomes are subverted to the +1 reading frame (ORF2) in a process stimulated by polyamines.

just upstream to it that programmes the repositioning of the ribosome. Frameshifting efficiency remains unaffected when the sequence of the stop codon is converted into the sequence of one of the other stop codons, but it is severely inhibited when the stop codon is replaced by a sense codon. The importance of the stop function is further emphasized by recent demonstrations that the yeast prion [PSI^+], which represents a non-functional aggregated conformation of the translational release factor eRF3, or inactivation of eRF3 by interaction with the interferon-induced RNAse L, increase frameshifting efficiency [11,12]. The mechanism by which polyamines stimulate frameshifting efficiency is still unknown.

Mechanism of ODC inactivation and stimulation of ODC degradation by Az

Az was originally identified as an ODC inhibitory activity that is stimulated by an increase in the intracellular polyamine concentration [13]. Upon its cloning, it was demonstrated that the affinity of Az for ODC subunits is significantly higher than the affinity of ODC subunits for each other [14]. This, together with the rapid and frequent association and dissociation of ODC subunits, enables efficient trapping of transient ODC monomers by Az, forming a tightly bound heterodimer. The strong association of ODC monomers with Az prevents their re-association to form active ODC homodimers. Although ODC inactivation is a clear function of Az, in most cases it is only an intermediate step in the process of targeting ODC subunits to ubiquitin-independent degradation by the 26S proteasome [15] (Figure 2). Interaction with Az seems to impose conformational alteration

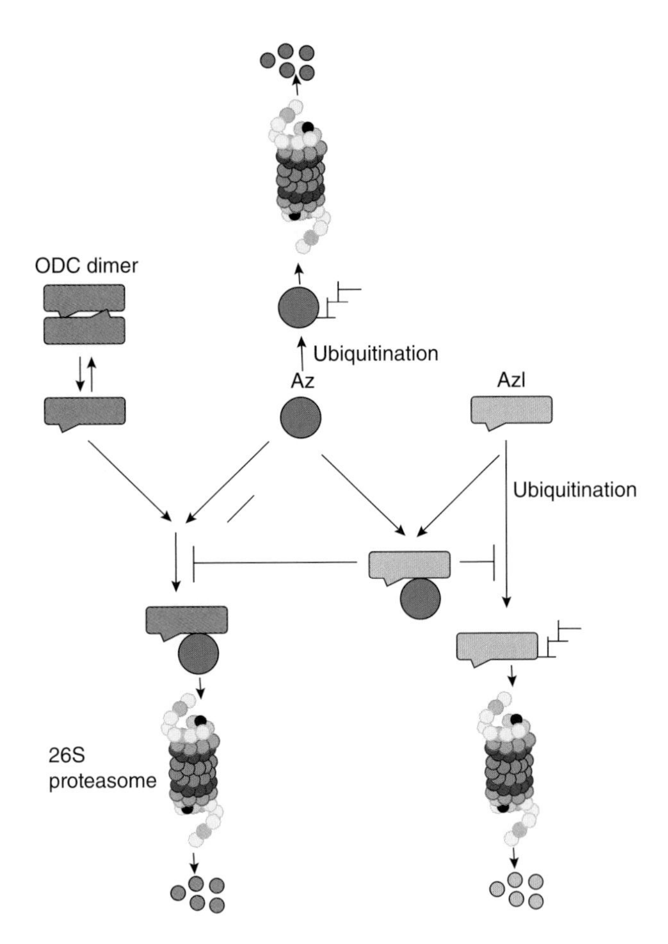

Figure 2. Regulatory interactions between ODC, Az and AzI
ODC exists in equilibrium between its active dimeric form and inactive monomers. Az, whose synthesis is regulated by a polyamine-stimulated ribosomal frameshifting mechanism, traps transient ODC monomers and brings them to the 26S proteasome for degradation. Az is not degraded together with ODC, but it is recycled to support additional rounds of ODC degradation. Independently of ODC, Az is also subjected to proteasomal degradation. Its degradation appears to require ubiquitination. The ability of Az to regulate ODC degradation is negated by the ODC-related enzymatically inactive protein AzI, whose monomeric existence contributes to its high affinity to Az. Owing to this high affinity, which is greater than the affinity of ODC monomers for Az, AzI maintains Az in a stable complex, thus preventing Az from fulfilling its regulatory cellular roles. AzI is also a rapidly degraded protein, but, despite its similarity to ODC, it is degraded in a ubiquitin-dependent manner. Interaction with Az stabilizes AzI by inhibiting its ubiquitination.

on ODC, resulting in the exposure of an ODC segment encompassing its most C-terminal 37 amino acids. This C-terminal segment functions as the proteasome recognition signal [16]. Exposure of the C-terminal targeting signal enhances the interaction of ODC with the proteasome without stimulating proteasomal activity. Another segment of ODC, encompassing amino acids 117–140, is also essential for stimulating ODC degradation, since it is required for binding to Az [16]. *Trypanosoma brucei* ODC, which is a stable protein, lacks sequences that parallel the C-terminal segment of the mammalian enzyme and it does not bind Az [17,18]. Although refractory to Az binding, trypanosome ODC is converted into a rapidly degraded protein when the C-terminal segment of the mammalian enzyme is appended to its C-terminus, probably because, in the context of the trypanosome enzyme, the C-terminal mammalian segment is exposed without requiring interaction with Az. Interestingly, although, like the trypanosome enzyme, yeast ODC also lacks a C-terminal targeting segment, it is rapidly degraded in yeast cells in an Az-dependent manner [19]. It appears that yeast ODC is targeted to degradation by an N-terminal segment that can be replaced by the C-terminal targeting segment of the mammalian enzyme. At present, it is not known whether, like the C-terminal segment of the mammalian enzyme, the N-terminal yeast segment can also function as a dominant transferrable degradation signal.

Segments of Az that affect its ability to stimulate ODC degradation

Az also has two segments that are required for its ability to stimulate ODC degradation [20]. A large segment encompassing the C-terminal half of the molecule is important for the ability to bind ODC. Although the binding mediated by this sequence is sufficient to inactivate ODC, it is insufficient for stimulating ODC degradation. A smaller N-terminal segment is required to induce ODC degradation. However, the mechanism by which this sequence affects the degradation process is still unknown.

The structure of Az in solution was determined using NMR methods [21]. Az was found to contain eight β-strands and two α-helices, with the strands forming a mixture of parallel and antiparallel β-sheets. A significant proportion of the residues that are highly conserved among Azs were found on the surface. Some of these residues form a negatively charged patch that might interact with an electropositive surface located in the putative Az-binding site of ODC.

Az inhibits polyamine uptake

In addition to regulating ODC activity and degradation, Az inhibits uptake of polyamines and stimulates their excretion [22] (Figure 3). The mechanism by which Az regulates transport of polyamines across the plasma membrane is mostly unknown, as is the basic process of polyamine transport in eukaryotic cells. Interestingly, Az has only a minor effect on polyamine uptake in yeast cells.

Az targets additional proteins that are unrelated to the polyamine metabolism for degradation

Although components of polyamine metabolism are the natural targets of Az, a number of studies suggested that Az might also target for degradation proteins that do not belong to the polyamine metabolic pathway or to its regulation (for a review, see [23]). It was suggested that Az, together with the proteasome β subunit HsN3, mediates the targeting of the signal transducer Smad1 to the proteasome [24,25]. Az was also demonstrated to interact and

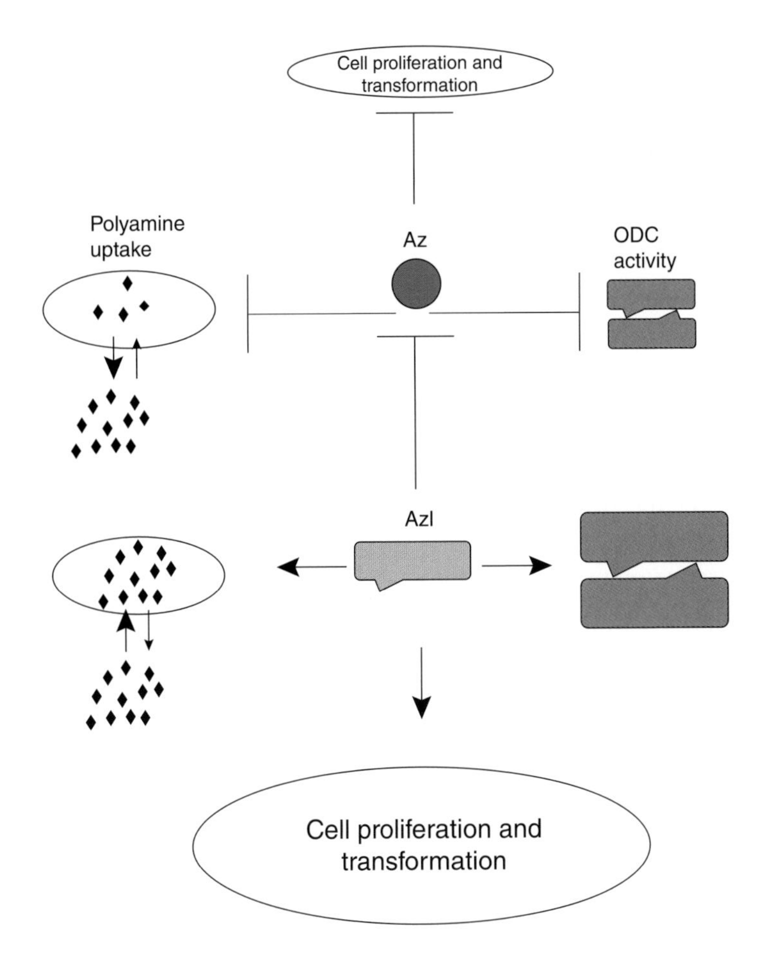

Figure 3. AzI controls the ability of Az to regulate cellular polyamines and cellular proliferation
Az inhibits ODC activity predominantly by targeting ODC to proteosomal ubiquitin-independent degradation. In addition, Az inhibits polyamine uptake and stimulates polyamine excretion via an as yet unknown mechanism. These effects of Az result in inhibition of cellular proliferation and sometimes even in the induction of apoptotic cell death. AzI, which binds Az with greater affinity than ODC, neutralizes Az, resulting in increased ODC activity and polyamine uptake, and thereby in an increase in cellular proliferation and tumorigenesis.

stimulate the degradation of two growth-related proteins, cyclin D1 and Aurora A [26,27]. Since, in contrast with ODC, whose degradation is always ubiquitin-independent, these proteins are also degraded in a ubiquitin-dependent manner, it is not clear when and under what physiological conditions each of these degradation pathways operates.

Az is a rapidly degraded protein

It is widely accepted that Az acts catalytically; namely, it is recycled to support additional rounds of ODC degradation [23]. Compatible with a catalytic mode of action, Az is not degraded together with ODC when presenting ODC to the proteasome (Figure 2). Nevertheless, Az is a rapidly degraded protein [28]. Interestingly, the degradation of Az is ubiquitin-dependent [28], and in yeast it is inhibited by polyamines. It is unclear how Az escapes degradation while escorting ODC to the proteasome, and how it is recognized as a substrate for ubiquitination.

Forced Az overexpression interferes with cellular proliferation and tumour development

The ability of Az to stimulate degradation of ODC and possibly of other growth-regulating proteins, together with its ability to inhibit polyamine uptake, strongly suggest that Az may inhibit cellular proliferation (Figure 3). Therefore it is not surprising that Az acts as a negative regulator of cellular proliferation and of tumour development both when expressed in cultured cells and in transgenic mice [29,30]. In some experimental systems, artificial Az overexpression provokes apoptotic cell death. It is therefore not surprising that in all studies performed in cultured cells, Az was expressed in an inducible manner. When expressed in transgenic animals, Az interferes with tumour development.

A family of Az proteins

The Az species described above, termed Az1, appears to be the prototype of a family of Azs containing three members characterized to date. A second member of this family, termed Az2, is extremely interesting as it has a tissue distribution similar to that of Az1, but it is expressed at much lower levels [31]. Az2 is more conserved evolutionarily than Az1, suggesting that it is likely to have a significant biological role. Its minority co-existence with Az1 suggests that its role might be different from that of Az1. In support of this possibility, is a recent demonstration that the level of Az2, but not of Az1, changes in neuroblastomas [32]. Although Az2 inhibits ODC activity, stimulates its degradation in cells and inhibits polyamine uptake as efficiently as Az1, it fails to promote ODC degradation in an *in vitro* reaction. The reasons and the significance of this odd behaviour are presently unknown. Also, it is presently unknown whether Az2 might possess additional functions not shared by Az1.

A third member of the antizyme family, Az3, is unique in being tissue-specific, expressed predominantly in testis. It is observed only in haploid germinal cells [33,34]. Interestingly, the localization of Az3 is different from that of ODC, and it is found mainly in the outer part of the seminiferous tubules where spermatogonia and spermatocytes are located [35]. This is compatible with the observation that Az3 does not target ODC to degradation and therefore might have a different role. Such an alternative role may be reflected by its interaction with the germ-cell-specific protein, gametogenetin protein-1. However, the functional consequence for this interaction has not yet been demonstrated.

Subcellular localization of Az1 is determined by alternative utilization of initiation codons

As mentioned above, *Az1* mRNA contains two AUG codons that can initiate translation. Although translation generally starts at the first AUG codon, translation of *Az1* is initiated predominantly at the second initiation codon because it is situated within a superior sequence context for translation initiation. Nevertheless, some initiations occur at the first AUG as well, and the products of the two initiation events are observed both *in vivo* and *in vitro* as 24.5 and 29 kDa isoforms. The segment located between the two initiation codons contains a positive amphiphathic helix that is part of a mitochondrial localization signal [36,37]. Thus only the minor long form is localized to the mitochondria in transfected cells and imported into mitochondria in an *in vitro* assay [36]. Although both forms stimulate ODC degradation and inhibit polyamine uptake through the plasma membrane, neither affects polyamine uptake by rat liver mitochondria [38]. Az1 also contains two independent nuclear export signals [39]. One is N-terminal, overlapping the mitochondrial localization signal, whereas the other resides in the central part of the protein. The relevance of the central signal is uncertain as it is functional only in the context of an N-terminal truncation. In agreement with the existence of such signals, shuttling of Az between the cytoplasm and nucleus has been documented [39–41].

AzI

AzI was originally described as an activity capable of inhibiting Az functions [42]. Although it is highly homologous with ODC, AzI is a distinct protein that lacks ornithine-decarboxylating activity and inhibits all members of the Az family. Recently it was demonstrated that expression of AzI might be repressed by polyamines via utilization of upstream ORFs initiating at non-canonical initiation codons [43].

Structural features explain the ability of AzI to negate Az functions

Recent studies have demonstrated that, to a large extent, the ability of AzI to inhibit Az is an outcome of its structure [44]. Although, like ODC, AzI crystallizes as a dimer, fewer interactions at the dimer interface, a smaller buried surface area, and lack of symmetry of the interactions between residues from the two monomers suggest that, under physiological conditions, AzI is actually a monomer [44]. Indeed, biochemical studies have confirmed that AzI exists as a monomer, whereas ODC is dimeric. As a monomer, AzI is more available for interaction with Az compared with ODC subunits that are in a constant state of association/dissociation. Since it is unlikely that greater availability is the only reason for this higher affinity, it is possible that the sequence of these two proteins may also be of importance for determining the affinity. Based on comparison between the sequence of mouse ODC, which binds Az, and trypanosome ODC, which does not bind Az, the segment encompassing amino acids 117–140 was suggested as the putative Az-binding site of mouse ODC [18]. However, based on comparison between the structures of human and trypanosome ODC, it was suggested that the Az-binding segment might actually be larger [45].

Since the active site of ODC is formed at the interface between the two monomers [46], the monomeric existence of AzI helps to explain why it lacks ornithine-decarboxylating activity. The observation that AzI is unable to bind PLP provides an additional and independent explanation for the lack of enzymatic activity [44].

AzI is degraded via the ubiquitin system

Like ODC, AzI is a rapidly degraded protein. However, in contrast with the ubiquitin-independent degradation of ODC, AzI is degraded in an Az-independent, ubiquitin-dependent manner [47]. In contrast with the interaction of Az with ODC, which greatly stimulates ODC degradation, interaction with Az actually stabilizes AzI by interfering with its ubiquitination. Since both partners of the Az–AzI complex are stabilized, it is tempting to suggest that AzI buffers Az by maintaining it in a stable complex. It will be of interest to identify conditions that might cause disintegration of this complex.

AzI stimulates cellular proliferation

Since AzI negates the ability of Az to stimulate ODC degradation and inhibit polyamine uptake, it is expected to elicit a growth-stimulatory effect (Figure 3). Indeed, several lines of evidence provide support to this notion (for a review, see [48]). *AzI* mRNA is rapidly induced in quiescent cells following growth stimulation by serum-derived growth factors. The induction of *AzI* mRNA precedes that of *ODC* mRNA, suggesting that AzI synthesis might protect ODC from Az. DNA array analysis has demonstrated increased levels of *AzI* mRNA in gastric tumours compared with nearby healthy tissue [49].

The human *AzI* gene is located on chromosome 8q22.3, and amplification of this region is associated with several tumours. Ectopic expression of AzI leads to increased proliferation and to cellular transformation. Conversely, silencing of AzI using siRNA (small interfering RNA) is associated with inhibition of ODC activity and with reduced cell proliferation. However, it is unclear whether the growth-promoting role of AzI is executed only through manipulating polyamine metabolism, since AzI has been demonstrated to stabilize cyclin D1 even in the absence of its Az-binding segment [50]. The most compelling evidence demonstrating that AzI has a true and meaningful physiological role was recently provided through the demonstration that mice lacking functional AzI die at birth, exhibiting abnormal liver morphology that is accompanied by increased ODC degradation and perturbed biosynthesis of putrescine and spermidine [51].

AzI variants

Accumulating evidence obtained mainly from databases suggests the existence of AzI isoforms that result from differences in coding and non-coding regions, resulting from alternative splicing and differential utilization of polyadenylation signals. Although in some of these isoforms the coding region remains unaffected, there are isoforms that exhibit differences in the coding region, as well. It is presently unknown whether the isoforms with alterations in the coding region are able to regulate polyamine metabolism, or whether they are involved in regulating other cellular processes.

AzI-2

Recent studies have demonstrated that mammalian cells contain another type of AzI, termed ODCp (ODC paralogue) or AzI-2 [52]. AzI-2 is expressed only in brain and testis [35,52] and, like AzI-1, it lacks ODC activity. Like AzI-1, AzI-2 is also rapidly degraded in a ubiquitin-dependent manner and, when artificially overexpressed, it stimulates ODC activity, polyamine uptake and cellular proliferation, although less efficiently than AzI-1 [53,54]. The spatial expression pattern of AzI-2 is similar to that of Az3, both being expressed in the haploid germinal cells. In addition, AzI-2 and Az3 are expressed at minimal levels during the first 3 postnatal weeks, and are highly induced at the fourth week [35]. These results suggest that AzI-2 and Az3 may have a role in spermiogenesis.

Conclusions

Although significant progress has been made in our understanding of the regulation of the cellular polyamine metabolism, there are still aspects that require additional clarification. These include (i) the characterization of the physical interactions of Az with ODC and AzI; (ii) the role of the N-terminal segment of Az, which enables its ability to degrade ODC; (iii) characterization of the specific components of the ubiquitin system that mediate the degradation of Az and AzI; and (iv) determination of their possible regulation by polyamines. A major task will be the detailed characterization of the process of polyamine transport across the plasma membrane and the role Az plays in regulating this process.

Because of the importance of polyamines for cellular growth, polyamine metabolism has become a target of therapeutic efforts in battling cancer and other hyperproliferative diseases. Despite initial enthusiasm, inhibitors of key enzymes in the polyamine biosynthesis pathway were found to be rather ineffective owing to the ability of cells to overcome their effect by accumulating polyamines from their environment. This dominant effect of the uptake activity suggested the therapeutic use of toxic structural analogues of polyamines that cross the plasma membrane through the polyamine uptake system. Since the effectiveness of such toxic polyamine analogues also depends on the level of AzI, which negates Az function in the treated cells, an additional challenge will be to match optimal treatment, namely the choice of synthesis inhibitors versus polyamine analogues or their combination, to the composition of the molecular players, especially ODC and AzI, in the treated cells.

Summary

- *ODC, the first rate-limiting enzyme in the biosynthesis of polyamines, is characterized by an extremely rapid degradation rate.*
- *ODC is degraded in a ubiquitin-independent manner.*
- *The degradation of ODC is mediated by a polyamine-induced protein termed Az that traps transient ODC monomers preventing their reassociation to form active dimers, and targets them to ubiquitin-independent degradation by the 26S proteasome.*
- *In addition to stimulating ODC degradation, Az inhibits uptake and stimulates excretion of polyamines via an, as yet unresolved, mechanism.*
- *In addition to affecting ODC degradation and polyamine transport, Az stimulates the degradation of other growth-regulating proteins.*
- *Forced Az overexpression inhibits cellular proliferation, tumour development and, in some systems, leads to apoptotic death.*
- *Az is regulated by an inactive ODC-related protein, termed AzI, which binds Az with high affinity and keeps it in a stable complex.*
- *Like ODC, Az and AzI are rapidly degraded proteins but, unlike ODC, their degradation is ubiquitin-dependent.*
- *Az is not degraded together with ODC while presenting it to the 26S proteasome.*
- *In contrast with its ability to stimulate ODC degradation, interaction with Az stabilizes AzI by inhibiting its ubiquitination.*
- *Forced AzI overexpression negates Az functions, and therefore stimulates cellular proliferation and transformation.*

Funding

Research in the Kahana laboratory on the regulation of polyamine metabolism is supported by grants from the Israel Academy of Science and Humanities, The Leo and Julia Forchheimer Center for Molecular Genetics, The M.D. Moross Institute for Cancer Research and The Y. Leon Benoziyo Institute for Molecular Medicine at the Weizmann Institute of Science. C.K is the incumbent of the Jules J. Mallon Professorial chair in Biochemistry.

References

1. Park, M.H., Cooper, H.L. and Folk, J.E. (1981) Identification of hypusine, an unusual amino acid, in a protein from human lymphocytes and of spermidine as its biosynthetic precursor. Proc. Natl. Acad. Sci. U.S.A. **78**, 2869–2873
2. Fakler, B., Brandle, U., Bond, C., Glowatzki, E., Konig, C., Adelman, J.P., Zenner, H.P. and Ruppersberg, J.P. (1994) A structural determinant of differential sensitivity of cloned inward rectifier K^+ channels to intracellular spermine. FEBS Lett. **356**, 199–203

3. Haghighi, A.P. and Cooper, E. (1998) Neuronal nicotinic acetylcholine receptors are blocked by intracellular spermine in a voltage-dependent manner. J. Neurosci. **18**, 4050–4062
4. Ransom, R.W. and Stec, N.L. (1988) Cooperative modulation of [^{3}H]MK-801 binding to the *N*-methyl-D-aspartate receptor-ion channel complex by L-glutamate, glycine, and polyamines. J. Neurochem. **51**, 830–836
5. Scott, R.H., Sutton, K.G. and Dolphin, A.C. (1993) Interactions of polyamines with neuronal ion channels. Trends Neurosci. **16**, 153–160
6. Southan, G.J., Szabo, C. and Thiemermann, C. (1994) Inhibition of the induction of nitric oxide synthase by spermine is modulated by aldehyde dehydrogenase. Biochem. Biophys. Res. Commun. **203**, 1638–1644
7. Casero, Jr, R.A. and Marton, L.J. (2007) Targeting polyamine metabolism and function in cancer and other hyperproliferative diseases. Nat. Rev. Drug Discov. **6**, 373–390
8. Pegg, A.E. (2006) Regulation of ornithine decarboxylase. J. Biol. Chem. **281**, 14529–14532
9. Matsufuji, S., Matsufuji, T., Miyazaki, Y., Murakami, Y., Atkins, J.F., Gesteland, R.F. and Hayashi, S. (1995) Autoregulatory frameshifting in decoding mammalian ornithine decarboxylase antizyme. Cell **80**, 51–60
10. Rom, E. and Kahana, C. (1994) Polyamines regulate the expression of ornithine decarboxylase antizyme *in vitro* by inducing ribosomal frame-shifting. Proc. Natl. Acad. Sci. U.S.A. **91**, 3959–3963
11. Le Roy, F., Salehzada, T., Bisbal, C., Dougherty, J.P. and Peltz, S.W. (2005) A newly discovered function for RNase L in regulating translation termination. Nat. Struct. Mol. Biol. **12**, 505–512
12. Namy, O., Galopier, A., Martini, C., Matsufuji, S., Fabret, C. and Rousset, J.P. (2008) Epigenetic control of polyamines by the prion [PSI(+)]. Nat. Cell Biol. **10**, 1069–1075
13. Fong, W.F., Heller, J.S. and Canellakis, E.S. (1976) The appearance of an ornithine decarboxylase inhibitory protein upon the addition of putrescine to cell cultures. Biochim. Biophys. Acta **428**, 456–465
14. Miyazaki, Y., Matsufuji, S. and Hayashi, S. (1992) Cloning and characterization of a rat gene encoding ornithine decarboxylase antizyme. Gene **113**, 191–197
15. Murakami, Y., Matsufuji, S., Kameji, T., Hayashi, S., Igarashi, K., Tamura, T., Tanaka, K. and Ichihara, A. (1992) Ornithine decarboxylase is degraded by the 26S proteasome without ubiquitination. Nature **360**, 597–599
16. Coffino, P. (2001) Regulation of cellular polyamines by antizyme. Nat. Rev. Mol. Cell Biol. **2**, 188–194
17. Ghoda, L., Phillips, M.A., Bass, K.E., Wang, C.C. and Coffino, P. (1990) Trypanosome ornithine decarboxylase is stable because it lacks sequences found in the carboxyl terminus of the mouse enzyme which target the latter for intracellular degradation. J. Biol. Chem. **265**, 11823–11826
18. Li, X. and Coffino, P. (1992) Regulated degradation of ornithine decarboxylase requires interaction with the polyamine-inducible protein antizyme. Mol. Cell. Biol. **12**, 3556–3562
19. Gandre, S. and Kahana, C. (2002) Degradation of ornithine decarboxylase in *Saccharomyces cerevisiae* is ubiquitin independent. Biochem. Biophys. Res. Commun. **293**, 139–144
20. Li, X. and Coffino, P. (1994) Distinct domains of antizyme required for binding and proteolysis of ornithine decarboxylase. Mol. Cell. Biol. **14**, 87–92
21. Hoffman, D.W., Carroll, D., Martinez, N. and Hackert, M.L. (2005) Solution structure of a conserved domain of antizyme: a protein regulator of polyamines. Biochemistry **44**, 11777–11785
22. Mitchell, J.L., Judd, G.G., Bareyal-Leyser, A. and Ling, S.Y. (1994) Feedback repression of polyamine transport is mediated by antizyme in mammalian tissue-culture cells. Biochem. J. **299**, 19–22
23. Mangold, U. (2005) The antizyme family: polyamines and beyond. IUBMB Life **57**, 671–676
24. Gruendler, C., Lin, Y., Farley, J. and Wang, T. (2001). Proteasomal degradation of Smad1 induced by bone morphogenetic proteins. J. Biol. Chem. **276**, 46533–46543
25. Lin, Y., Martin, J., Gruendler, C., Farley, J., Meng, X., Li, B.Y., Lechleider, R., Huff, C., Kim, R.H., Grasser, W.A. et al. (2002) A novel link between the proteasome pathway and the signal transduction pathway of the bone morphogenetic proteins (BMPs). BMC Cell Biol. **3**, 15
26. Lim, S.K. and Gopalan, G. (2007) Antizyme1 mediates AURKAIP1-dependent degradation of Aurora-A. Oncogene **26**, 6593–6603

27. Newman, R.M., Mobascher, A., Mangold, U., Koike, C., Diah, S., Schmidt, M., Finley, D. and Zetter, B.R. (2004) Antizyme targets cyclin D1 for degradation. A novel mechanism for cell growth repression. J. Biol. Chem. **279**, 41504–41511
28. Gandre, S., Bercovich, Z. and Kahana, C. (2002) Ornithine decarboxylase-antizyme is rapidly degraded through a mechanism that requires functional ubiquitin-dependent proteolytic activity. Eur. J. Biochem. **269**, 1316–1322
29. Fong, L.Y., Feith, D.J. and Pegg, A.E. (2003) Antizyme overexpression in transgenic mice reduces cell proliferation, increases apoptosis, and reduces *N*-nitrosomethylbenzylamine-induced forestomach carcinogenesis. Cancer Res. **63**, 3945–3954
30. Murakami, Y., Matsufuji, S., Miyazaki, Y. and Hayashi, S. (1994) Forced expression of antizyme abolishes ornithine decarboxylase activity, suppresses cellular levels of polyamines and inhibits cell growth. Biochem. J. **304**, 183–187
31. Ivanov, I.P., Gesteland, R.F. and Atkins, J.F. (1998) A second mammalian antizyme: conservation of programmed ribosomal frameshifting. Genomics **52**, 119–129
32. Hogarty, M.D., Norris, M.D., Davis, K., Liu, X., Evaeliou, N.F., Hayes, C.S., Pawel, B., Guo, R., Zhao, H., Sekyere, E., Keating J. et al. (2008) ODC1 is a critical determinant of MYCN oncogenesis and a therapeutic target in neuroblastoma. Cancer Res. **68**, 9735–9745
33. Ivanov, I.P., Rohrwasser, A., Terreros, D.A., Gesteland, R.F. and Atkins, J.F. (2000) Discovery of a spermatogenesis stage-specific ornithine decarboxylase antizyme: antizyme 3. Proc. Natl. Acad. Sci. U.S.A. **97**, 4808–4813
34. Tosaka, Y., Tanaka, H., Yano, Y., Masai, K., Nozaki, M., Yomogida, K., Otani, S., Nojima, H. and Nishimune, Y. (2000) Identification and characterization of testis specific ornithine decarboxylase antizyme (OAZ-t) gene: expression in haploid germ cells and polyamine-induced frameshifting. Genes Cells **5**, 265–276
35. Lopez-Contreras, A.J., Ramos-Molina, B., Martinez-de-la-Torre, M., Penafiel-Verdu, C., Puelles, L., Cremades, A. and Penafiel, R. (2009) Expression of antizyme inhibitor 2 in male haploid germinal cells suggests a role in spermiogenesis. Int. J. Biochem. Cell Biol. **41**, 1070–1078
36. Gandre, S., Bercovich, Z. and Kahana, C. (2003) Mitochondrial localization of antizyme is determined by context-dependent alternative utilization of two AUG initiation codons. Mitochondrion **2**, 245–256
37. Mitchell, J.L. and Judd, G.G. (1998) Antizyme modifications affecting polyamine homoeostasis. Biochem. Soc. Trans. **26**, 591–595
38. Hoshino, K., Momiyama, E., Yoshida, K., Nishimura, K., Sakai, S., Toida, T., Kashiwagi, K. and Igarashi, K. (2005) Polyamine transport by mammalian cells and mitochondria: role of antizyme and glycosaminoglycans. J. Biol. Chem. **280**, 42801–42808
39. Murai, N., Murakami, Y. and Matsufuji, S. (2003) Identification of nuclear export signals in antizyme-1. J. Biol. Chem. **278**, 44791–44798
40. Gritli-Linde, A., Nilsson, J., Bohlooly, Y.M., Heby, O. and Linde, A. (2001) Nuclear translocation of antizyme and expression of ornithine decarboxylase and antizyme are developmentally regulated. Dev. Dyn. **220**, 259–275
41. Schipper, R.G., Cuijpers, V.M., De Groot, L.H., Thio, M. and Verhofstad, A.A. (2004) Intracellular localization of ornithine decarboxylase and its regulatory protein, antizyme-1. J. Histochem. Cytochem. **52**, 1259–1266
42. Fujita, K., Murakami, Y. and Hayashi, S. (1982) A macromolecular inhibitor of the antizyme to ornithine decarboxylase. Biochem. J. **204**, 647–652
43. Ivanov, I.P., Loughran, G. and Atkins, J.F. (2008) uORFs with unusual translational start codons autoregulate expression of eukaryotic ornithine decarboxylase homologs. Proc. Natl. Acad. Sci. U.S.A. **105**, 10079–10084
44. Albeck, S., Dym, O., Unger, T., Snapir, Z., Bercovich, Z. and Kahana, C. (2008) Crystallographic and biochemical studies revealing the structural basis for antizyme inhibitor function. Protein Sci. **17**, 793–802
45. Almrud, J.J., Oliveira, M.A., Kern, A.D., Grishin, N.V., Phillips, M.A. and Hackert, M.L. (2000) Crystal structure of human ornithine decarboxylase at 2.1 Å resolution: structural insights to antizyme binding. J. Mol. Biol. **295**, 7–16

46. Tobias, K.E. and Kahana, C. (1993) Intersubunit location of the active site of mammalian ornithine decarboxylase as determined by hybridization of site-directed mutants. Biochemistry **32**, 5842–5847
47. Bercovich, Z. and Kahana, C. (2004) Degradation of antizyme inhibitor, an ornithine decarboxylase homologous protein, is ubiquitin-dependent and is inhibited by antizyme. J. Biol. Chem. **279**, 54097–54102
48. Mangold, U. (2006) Antizyme inhibitor: mysterious modulator of cell proliferation. Cell Mol. Life Sci. **63**, 2095–2101
49. Jung, M.H., Kim, S.C., Jeon, G.A., Kim, S.H., Kim, Y., Choi, K.S., Park, S.I., Joe, M.K. and Kimm, K. (2000) Identification of differentially expressed genes in normal and tumor human gastric tissue. Genomics **69**, 281–286
50. Kim, S.W., Mangold, U., Waghorne, C., Mobascher, A., Shantz, L., Banyard, J. and Zetter, B.R. (2006) Regulation of cell proliferation by the antizyme inhibitor: evidence for an antizyme-independent mechanism. J. Cell Sci. **119**, 2583–2591
51. Tang, H., Ariki, K., Ohkido, M., Murakami, Y., Matsufuji, S., Li, Z. and Yamamura, K. (2009) Role of ornithine decarboxylase antizyme inhibitor *in vivo*. Genes Cells **14**, 79–87
52. Pitkanen, L.T., Heiskala, M. and Andersson, L.C. (2001) Expression of a novel human ornithine decarboxylase-like protein in the central nervous system and testes. Biochem. Biophys. Res. Commun. **287**, 1051–1057
53. Lopez-Contreras, A.J., Ramos-Molina, B., Cremades, A. and Penafiel, R. (2008) Antizyme inhibitor 2 (AZIN2/ODCp) stimulates polyamine uptake in mammalian cells. J. Biol. Chem. **283**, 20761–20769
54. Snapir, Z., Keren-Paz, A., Bercovich, Z. and Kahana, C. (2008) ODCp, a brain- and testis-specific ornithine decarboxylase paralogue, functions as an antizyme inhibitor, although less efficiently than AzI1. Biochem. J. **410**, 613–619

Essays Biochem. (2009) **46**, 63–76; doi:10.1042/BSE0460005

5

Cells and polyamines do it cyclically

Kersti Alm and Stina Oredsson[1]

Department of Cell and Organism Biology, Lund University, Helgonavägen 3B, SE-223 62 Lund, Sweden

Abstract

Cell-cycle progression is a one-way journey where the cell grows in size to be able to divide into two equally sized daughter cells. The cell cycle is divided into distinct consecutive phases defined as G_1 (first gap), S (synthesis), G_2 (second gap) and M (mitosis). A non-proliferating cell, which has retained the ability to enter the cell cycle when it receives appropriate signals, is in G_0 phase, and cycling cells that do not receive proper signals leave the cell cycle from G_1 into G_0. One of the major events of the cell cycle is the duplication of DNA during S-phase. A group of molecules that are important for proper cell-cycle progression is the polyamines. Polyamine biosynthesis occurs cyclically during the cell cycle with peaks in activity in conjunction with the G_1/S transition and at the end of S-phase and during G_2-phase. The negative regulator of polyamine biosynthesis, antizyme, shows an inverse activity compared with the polyamine biosynthetic activity. The levels of the polyamines, putrescine, spermidine and spermine, double during the cell cycle and show a certain degree of cyclic variation in accordance with the biosynthetic activity. When cells in G_0/G_1-phase are seeded in the presence of compounds that prevent the cell-cycle-related increases in the polyamine pools, the S-phase of the first cell cycle is prolonged, whereas the other phases are initially unaffected. The results point to an important role for polyamines with regard to the ability of the cell to attain optimal rates of DNA replication.

[1]*To whom correspondence should be addressed (email Stina.Oredsson@cob.lu.se).*

Introduction

The fundamental unit of a multicellular organism, whether plant or animal, is the cell. The size of the organism depends on the number and size of cells. As all organisms in principle arise from one cell, the development of an organism depends on mechanisms whereby one cell can multiply into the number of cells that are required to build a specific organism. The basic cell growth and division cycle is called the cell cycle. The cell is an extremely complex structure composed of thousands of molecules of different sizes and with different roles in the life cycle of the cell. This chapter concerns the role for the polyamines in the cell cycle of mammalian cells.

Polyamines are present as essential components in all types of cells in multicellular organisms, and they are involved in several different cellular processes and structures. Polyamines are absolutely necessary for the growth of cells and tissues. The three main natural polyamines are putrescine, spermidine and spermine. They are aliphatic organic amines with various carbon chain lengths. At the pH in the cell, they are positively charged with the charge distributed along the entire length of the carbon chain. This distribution of the positive charge enables polyamines to interact in a specific way with polyanionic molecules in the cell, thus affecting their function. Extremely complicated processes involving biosynthesis, catabolism and transport over the cell membrane have evolved that maintain the cellular polyamine homoeostasis. With respect to polyamine homoeostasis during the cell cycle, biosynthesis has so far been mainly investigated. One means of trying to elucidate the role for polyamines during the cell cycle has been by using compounds that interfere with polyamine homoeostasis, resulting in decreased polyamine pools. Studies have also been performed with cell lines lacking biosynthetic enzymes. It has been determined that normal cell-cycle progression does indeed require certain levels of polyamines. However, not only polyamines are needed, but also an intricate web of proteins and enzymes, as well as other molecules. In this chapter, the cell cycle will first be described, together with some important cell-cycle regulatory mechanisms. The polyamines will then be introduced into this context and changes in polyamine homoeostasis during the cell cycle will be presented. Finally, cell-cycle effects caused by polyamine-pool depletion will be discussed as well as future research directions. This overview focuses on effects detected within one cell cycle of treatment with compounds that cause changes in the polyamine pools, something that we define as early cell-cycle effects. Early effects of polyamine-pool depletion may provide better evidence for the direct role for polyamines in cell-cycle regulation.

The cell cycle

The process by which one cell divides into two is fundamental to all growth, whether it is in connection with the developmental growth of an embryo into a young functional organism, further development into an adult, replacement

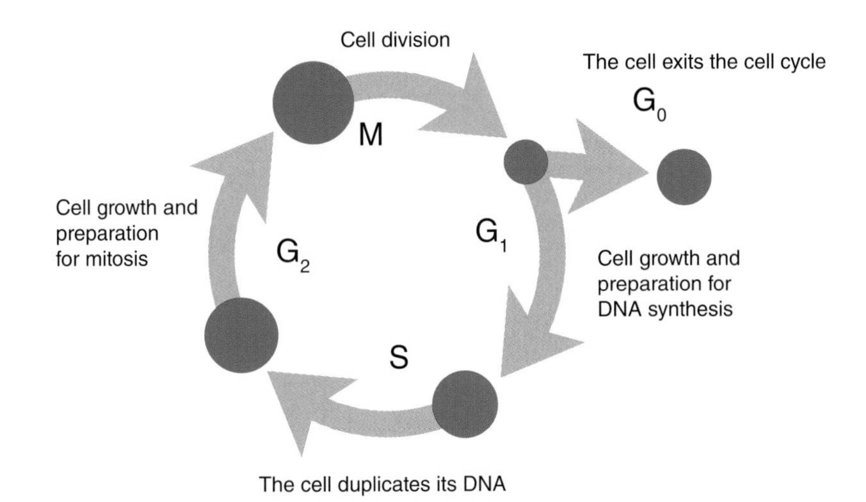

Figure 1. A schematic representation of the cell cycle

of injured cells in any life phase of the organism or uncontrolled growth of a cancer cell. The quiescent non-proliferating cell is said to be in a G_0 phase of its life cycle (Figure 1). Unlike terminally differentiated cells, G_0 cells retain the ability to enter the proliferative cell cycle as a response to extracellular factors. The purpose of the active cell cycle is to double the structural elements and functional capacities of the cell so that it can divide into two equal daughter cells at the end of the process. Thus the cell cycle includes two major processes: cell growth and cell division. The cell growth process is divided into three consecutive, biologically defined phases: G_1 (first gap), S (DNA synthesis) and G_2 (second gap), which are followed by the division phase called M (mitosis) (Figure 1). Completion of a specific sequence of metabolic events in each phase enables the cell to proceed to the next phase. The key event of the cell-growth process is the duplication of DNA taking place in S-phase.

Starting from a quiescent state in G_0, the progression of cells into G_1 depends absolutely on the availability of specific growth factors and a suitable extracellular environment (Figure 2). Quiescent cells initially require growth factors to initiate early regulatory processes referred to as competence [1]. There are a number of different growth factors, such as PDGF (platelet-derived growth factor) and EGF (epidermal growth factor) [2–4]. Different cells are stimulated by different growth factors depending on which receptors they display on their surface. Binding of a growth factor to its receptor initiates immediate effects on the activation of signal transduction pathways [e.g. the MAPK (mitogen-activated protein kinase) pathway], resulting in activation of immediate early genes (also known as primary response genes) (Figure 2). Transcription of *FOS* and *JUN* has been shown in numerous model systems to be induced within minutes of growth-factor stimulation of cells in G_0 as a

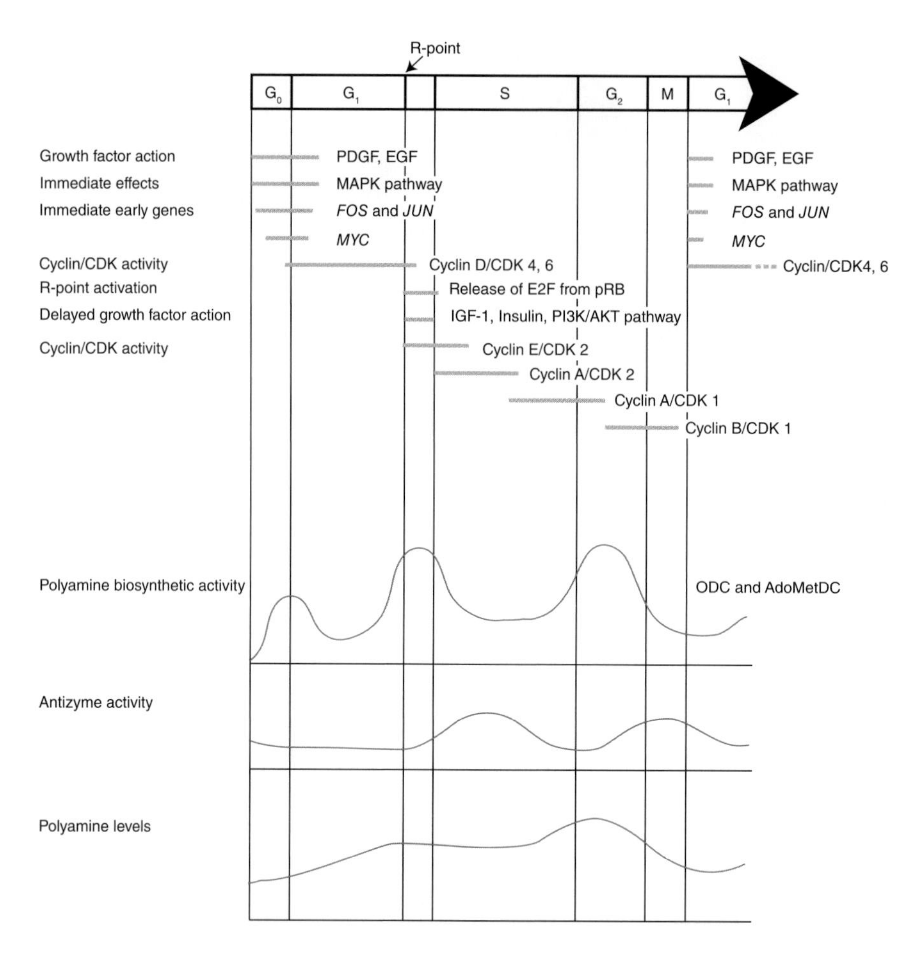

Figure 2. A schematic representation of the timing of factors involved in the regulation of the cell cycle
Italic font indicates gene expression; roman font indicates protein expression. Light blue lines indicate the timing of gene or protein expression.

consequence of growth-factor action. Another immediate early gene is *MYC*, which is induced slightly after *JUN* and *FOS* (Figure 2). The rapid protein synthesis-independent induction of immediate early genes is followed by the subsequent protein synthesis-dependent induction of secondary response genes. The induction of secondary response genes is distinct from that of primary response genes in requiring *de novo* protein synthesis, i.e. cycloheximide treatment inhibits the accumulation of their respective mRNA. Thus the generally accepted model of growth-factor-induced gene expression has two major components: the initial induction of immediate early genes, followed by a compulsory delay allowing translation of their mRNAs to produce the transcription factors that then induce the secondary response genes. However, there are also newer results indicating that gene induction can be divided into three steps: the immediate early, the delayed primary and the secondary response, where the first two steps do not require protein synthesis. The D-type cyclins belong to the secondary response genes and have a major role during G_1 progression (Figure 2).

D-type cyclins regulate the kinase function of CDK (cyclin-dependent kinase) 4 and 6 (Figure 2). Cyclin D/CDK4,6 complexes phosphorylate different proteins, thereby driving the cell through the G_1-phase, during which very rapid protein synthesis takes place. The cell eventually reaches the R-point (restriction point) (Figure 2) [5]. Once the cell passes the R-point it is committed and has to complete a full cell cycle [1]. At the R-point, the E2F transcription factors are released from the pRB (retinoblastoma protein) and related proteins p107 and p130. E2F are transcription factors required for the transcription of genes involved directly or indirectly in DNA replication [6]. Cells in G_0/G_1 express non-phosphorylated forms of pRB. Cyclin D/CDK4,6 complexes initiate the phosphorylation of pRB [7], whereas cyclin E, together with CDK2, orchestrate continued phosphorylation of pRB, resulting in its inactivation which leads to release of E2F. The release of E2F facilitates the activation of genes critical for S-phase progression.

In the presence of IGF-1 (insulin-like growth factor-1) or insulin, the cell will enter the S-phase and commence DNA synthesis without any additional requirements for growth factors (Figure 2) [8]. In the absence of IGF-1 or insulin, the cell will retain the ability to enter into S-phase for a short while; however, if the hormone peptides are not provided, the cell will exit the S-phase and enter the G_0-phase [9]. Binding of IGF-1 to the IGF-1 receptor activates, e.g. the PI3K (phosphoinositide 3-kinase)/Akt pathway, which is an important survival stimulation pathway necessary for the completion of the cell cycle (Figure 2).

As the cell enters and proceeds through S-phase, cyclin E/CDK2 activity rapidly diminishes and instead cyclin A exerts the activating function of CDK2 (Figure 2). In the middle of S-phase, cyclin A switches partner to CDK1, which in turn switches partner to cyclin B during the G_2-phase. The different cyclin/CDK complexes phosphorylate different proteins consecutively, and in that way irreversibly drive the cell cycle towards cell division.

Working against the cell-cycle stimulatory effects of CDKs are the CDK inhibitory proteins $p15^{INK4B}$, $p16^{INK4A}$, $p19^{INK4D}$, $p21^{WAF1/CIP1}$, $p27^{KIP1}$ and $p57^{KIP2}$ [7,10]. A number of phosphatases are also involved in modulating cell-cycle progression [11]. There are also checkpoint controls, besides the R-point, where the cell can be halted in the cell cycle if different kinds of damage are sensed. Damage induced in DNA after exposure of cells to ionizing radiation activates checkpoint pathways that inhibit progression of cells through the G_1- and G_2-phases, and induce a transient delay in the progression through S-phase [12].

At the end of mitosis, the cell divides into two daughter cells that continue directly into the G_1-phase if the necessary growth factors are provided. There seem to be certain differences in the molecular events at the G_0/G_1 transition compared with the M/G_1 transition, mainly resulting in prolonged transition before the commencement of S-phase in the former transition.

Polyamines and the cell cycle

After having described the cell cycle and some of the mechanisms that govern cell-cycle progression, changes in polyamine metabolism and polyamine levels that have been observed during the cell cycle will be added into this process. Possible functions of the polyamines during the cell cycle will be discussed in the next section in relation to the cell-cycle changes that follow polyamine-pool manipulation.

A cell in G_0, which is not proliferating, contains lower polyamine levels than when it is progressing through the cell cycle. The sizes of the polyamine pools are determined by biosynthesis, catabolism and uptake over the cell membrane. The increase in polyamine pools that take place when a cell is stimulated to proliferate is mainly due to activation of biosynthesis. Most work on polyamine biosynthesis during the cell cycle has been performed on ODC (ornithine decarboxylase) and AdoMetDC (*S*-adenosylmethionine decarboxylase).

ODC and AdoMetDC activities are very low in quiescent G_0 cells [13]. The ODC activity increases rapidly when quiescent G_0 cells are stimulated with serum (Figure 2) [14,15]. The ODC and AdoMetDC activities display two peaks during the G_1-to-M progression, one in conjunction with the G_1/S transition and the second in conjunction with the S/G_2 transition and G_2-phase (Figure 2) [16,17]. Thus it seems that when quiescent cells are stimulated to proliferate, three enzyme activity peaks are seen during the first cell cycle, whereas only two are found in actively proliferating cells. This is, however, not totally clear and needs further investigation.

ODC and *AdoMetDC* are supposed to belong to the group of secondary response genes, as cycloheximide treatment inhibits the massive accumulation of their mRNA after serum stimulation [15]. Actinomycin D treatment inhibits this mRNA accumulation as well. There is, however, evidence that there is a very low level of stable *ODC* mRNA in quiescent cells, which provides the basis for the peak in activity found early after growth stimulation of G_0 cells [18]. Hogan [18] showed that this early peak in ODC activity did not depend on mRNA synthesis, but only on protein synthesis. Another study has shown that the peak in ODC activity, found when quiescent G_0 cells were stimulated to proliferate by serum, appeared to be regulated differently than the first ODC peak in actively cycling cells [14]. The importance of this early peak after growth stimulation may give rise to speculation. It is well-known that *c*-myc is a transcription factor for ODC, and that *MYC* belongs to the immediate early genes [19]. However, polyamines are also known to participate in a positive-feedback loop in the regulation of the *MYC* gene [20]. Thus this early peak in ODC activity may be necessary for an optimal activation of *MYC* to ensure an optimal G_0/G_1 transition and optimal *c*-myc-induced gene transcription (Figure 3).

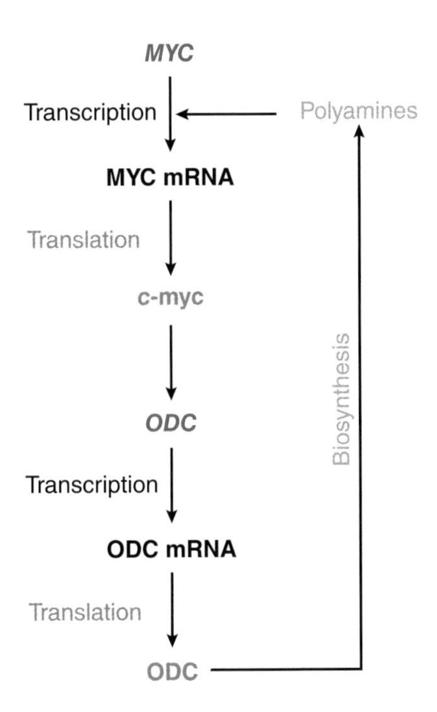

Figure 3. Role for polyamines in a positive-feedback loop in the regulation of the *MYC* gene
Dark blue text, gene; bold black text, mRNA; bold grey text, protein.

A general phenomenon of proliferating cells appears to be the peak in ODC and AdoMetDC activities found in conjunction with the G_1/S transition [17,21–23]. The temporal correlation between these activities and the onset of DNA replication suggests that polyamines are important for processes during S-phase, something that will be discussed below. The activity peak found during late S-phase and G_2-phase also seems to be a general phenomenon consistent with a role for the polyamines at the end of the cell cycle [17,22].

Several different mechanisms, such as the level of mRNA synthesis, the translational efficiency of mRNA and the stability of the enzyme, are part of the complex regulation of ODC activity [24,25]. One important regulating mechanism for enzyme stability involves antizyme, which induces ubiquitin-independent proteasomal degradation of ODC [26]. Antizyme activity was shown to fluctuate in an inverse correlation with ODC activity during the cell cycle, thus implicating a role for antizyme in the biphasic activity of ODC during the cell cycle (Figure 2) [27]. It has been suggested that the second peak in ODC activity depends on a cap-independent internal ribosome entry site in *ODC* mRNA that functions exclusively in the G_2/M-phase [28], a suggestion that has been contradicted by others [29].

Although polyamine biosynthetic activities show distinct biphasic changes during the cell cycle, such clear biphasic changes in polyamine levels have not always been found [17, 22, 30]. It is clear that polyamine pools increase during the cell cycle and in principle double in size from G_1- to the end of the G_2-phase (Figure 2); however, different studies show different kinetics in the changes of the pools [17, 22, 30].

The discussion above pertains to mammalian cells; however, similar biphasic changes in polyamine biosynthesis and polyamine levels have been found in synchronized tobacco BY-2 cells [31].

Polyamine-pool depletion and cell-cycle progression

Polyamine-pool depletion is achieved in cell lines and *in vivo* by treatment with compounds that inhibit polyamine biosynthesis and/or stimulate polyamine catabolism [32,33]. Since polyamine-pool depletion always results in inhibition of cell proliferation, many of these compounds are being exploited in the treatment of proliferative diseases such as cancer [32–34]. However, the stop in cell-cycle progression in cells treated with compounds that deplete the polyamine pool is not immediate, but instead the cell-cycle progression is gradually more and more affected. In general it takes several cell cycles before cell proliferation is totally halted. The early effects seen in the first cell cycle after the beginning of the treatment are described below in relation to the specific cell-cycle phases.

G_0/G_1 transition and G_1 progression

It has been shown that non-proliferating mouse Balb/c-3T3 fibroblasts in G_0 contained basal levels of polyamines that allowed an apparently normal G_0/G_1 transition, although polyamine biosynthesis was inhibited [35]. Even polyamine-depleted cells that were stimulated to proliferate by the addition of serum were able to enter the G_1-phase [35]. The actual G_0/G_1 transition was not investigated, but the time point at which cells entered S-phase was the same in control and inhibitor-treated cells. These studies were performed using the ODC inhibitor DFMO (2-difluoromethylornithine) [36]. Similar results were found when CHO (Chinese hamster ovary) cells were treated with DFMO [37]. DFMO treatment does not, however, lower the spermine pool [36]. Using the AdoMetDC inhibitor CGP48664 (4-amidinoindan-1-one 2′-amidinohydrazone) [38], it was shown that G_1-phase progression was not inhibited in CHO cells [39]. Treatment with CGP48664 decreased the spermidine and spermine pools, whereas the putrescine pool increased. When plateau-phase CHO cells were seeded in the presence of the polyamine analogue DENSPM (diethylnorspermine) [40], the progression to S-phase appeared to occur normally, although the levels of all three polyamines decreased [41].

These results and others [42] indicate that cells have a certain basal level of polyamines, which enable them to progress from G_0 through the G_0/G_1 transition and the G_1-phase of the first cell cycle, after seeding in the presence of compounds that inhibit polyamine biosynthesis (Figure 4).

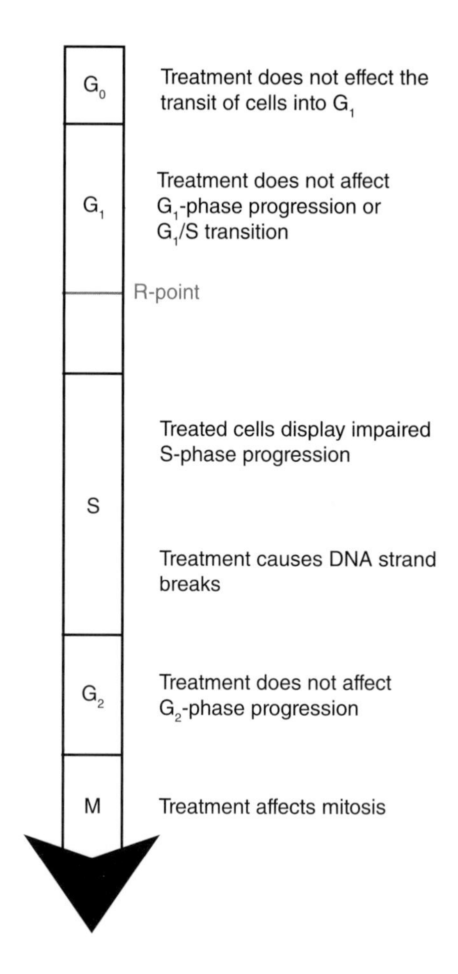

Figure 4. Cell-cycle kinetic effects during the first cell cycle after seeding cells in the presence of compounds that prevent the normal cyclic changes in polyamine biosynthesis
The blue line denotes the R-point.

G_1/S transition and S-phase progression

A large number of experiments with polyamine biosynthesis inhibitors have revealed an effect of polyamine depletion on the key event of the cell cycle, i.e. the replication of DNA [42]. Schaefer and Seidenfeld [35] showed that when non-proliferating mouse Balb/c-3T3 fibroblasts in G_0 were seeded in the presence of DFMO, S-phase progression was impaired after an apparently normal G_0 and G_1 progression [35]. Laitinen et al. [43] showed impaired S-phase progression during the first cell cycle after seeding CHO cells lacking ODC activity in medium without putrescine. They also showed that polyamine depletion did not significantly interfere with the nucleosomal organization of chromatin during S-phase. Using a bromodeoxyuridine DNA flow cytometric method, it was shown that the G_1/S transition was not affected after seeding cells in the presence of compounds that lower the polyamine pools, but that S-phase progression was impaired [37,39,41].

Thus it seems clear that the peak in polyamine biosynthesis occurring in conjunction with the G_1/S transition is important for a proper S-phase progression (Figures 2 and 4); however, the mechanism is not clear. Polyamines themselves may be important as immediate stabilizing molecules for newly synthesized DNA, as polyamines have shown important DNA-stabilizing functions [44]. A destabilized DNA may present a physical hindrance for effective DNA replication [45]. Using the single-cell gel electrophoresis assay, it was shown that DNA strand breaks were induced within one cell cycle after seeding cells in the presence of compounds that reduced the polyamine pools [46]. This would seem to confirm the DNA-stabilizing function of the polyamines.

It remains to be clarified whether it is only the decrease in polyamine pools that results in the delayed S-phase progression during the first cell cycle after seeding cells in the presence of compounds that deplete polyamines, or whether there are also other molecular changes as a consequence of low polyamine levels that may impinge on the replication machinery. One signal transduction pathway important for successful DNA replication is the PI3K/Akt pathway [47]. It has been shown that DFMO treatment for 24 h reduced the phosphorylation of Akt in neuroblastoma LAN-1 cells [48]. In another study, the phosphorylation of Akt was not affected in human chondrocytes treated with DENSPM for 24 h [49]. These seemingly contradictory results may only be a question of differences in polyamine-pool reduction. Further studies of how the PI3K/Akt pathway and other signal transduction pathways are affected by polyamine depletion before cells enter S-phase is a future arduous research project that is necessary to achieve an understanding of the molecular mechanisms behind the delay.

G_2- and M-phase progression

A second peak in polyamine biosynthesis takes place at the end of S-phase and in G_2-phase. However, the importance of this activity for G_2- and M-phase progression has been less well studied, presumably because no real overt effects have so far been found in the G_2- and M-phases during the first cell cycle after seeding cells in the presence of compounds that lower the polyamine pools. The study by Laitinen et al. [43] mentioned above, showed that polyamines may regulate S-phase progression without, however, affecting the nucleosomal organization of chromatin. They also found that the nucleosomal organization of chromatin was not affected in G_2 and that mitosis was not affected, but merely occurred later due to the prolonged S-phase. Similar results have been observed in breast cancer cell lines and a normal breast epithelial cell line treated with DENSPM [50,51], i.e. despite a prolongation of the S-phase, there was no effect on the length of the G_2+M-phase. This indicates that when a cell, seeded in the presence of a compound that lowers its polyamine pools, has traversed the S-phase, albeit at a reduced rate, there are no more negative changes caused by polyamine depletion during that first cell cycle (Figure 4). This notion clearly has to be investigated.

What happens during the second cell cycle and further on?

Cells seeded in the presence of compounds that deplete the polyamine pools will finally stop proliferating after one to several cell cycles, depending on the initial polyamine content and the concentration of the compound used. Many results indicate that the final stop point in the cell cycle depends on the status of the p53 gene [50–53]. In the presence of wild-type p53, cell-cycle progression is halted in the G_1-phase, whereas cells containing mutated p53 continue cycling; however, at a progressively slower rate with all cell-cycle phases being affected. It has been shown that the induction of cell-cycle arrest in cells containing wild-type p53 involves the p53/p21$^{WAF1/CIP1}$/pRb pathway [53], and an increase in p27^{KIP1} has also been found [54]. It also appears that the MAPK signal transduction pathway is involved in the induction of p53/ p21$^{WAF1/CIP1}$/pRb [55].

Conclusions

There is no doubt that the polyamines are important for the ability of the cell to keep up optimal rates of cell-cycle progression. Many results point to the S-phase being the most sensitive phase to polyamine depletion. The majority of the research of polyamines in cell-cycle regulation was carried out a number of years ago. With all the new knowledge of different cell-cycle regulatory systems, including signal transduction pathways, as well as a better knowledge of different mechanism of DNA replication, research into the mechanisms exerted by the polyamines in cell-cycle regulation should be intensified. Particularly, it would be a very important finding if any cell-cycle regulatory pathways could be elucidated as being responsible for the polyamine depletion-induced effects on the cell cycle.

Summary

- *Polyamine biosynthesis is activated when quiescent G_0 cells are stimulated by mitogens.*
- *Polyamine biosynthesis varies bicyclically during the active cell cycle, with one peak in conjunction with the G_1/S transition and a second at the end of S-phase and G_2-phase.*
- *When cells in G_0/G_1 are seeded in the presence of compounds that prevent the increases in polyamine pools, the S-phase of the first cell cycle is prolonged, whereas the other phases are unaffected.*
- *Polyamine depletion finally results in cell-cycle arrest after one to several cell cycles, depending on the efficiency of polyamine depletion.*

References

1. Pledger, W.J., Stiles, C.D., Antoniades, H.N. and Scher, C.D. (1977) Induction of DNA synthesis in BALB/c 3T3 cells by serum components: reevaluation of the commitment process. Proc. Natl. Acad. Sci., U.S.A. **74**, 4481–4485
2. Bravo, R. (1990) Genes induced during the G0/G1 transition in mouse fibroblasts. Semin. Cancer Biol. **1**, 37–46
3. Jones, S.M. and Kaslauskas, A. (2000) Connecting signaling and cell cycle progression in growth factor-stimulated cells. Oncogene **19**, 5558–5567
4. Reddy, G.P.V. (1994) Cell cycle: regulatory events in G→S transition of mammalian cells. J. Cell. Biochem. **54**, 379–386
5. Pardee, A.B. (1974) A restriction point for control of normal animal cell proliferation. Proc. Natl. Acad. Sci., U.S.A. **71**, 1286–1290
6. Neganova, I. and Lako, M. (2008) G1 to S phase cell cycle transition in somatic and embryonic stem cells. J. Anat. **213**, 30–44
7. Sherr, C.J. and Roberts, J.M. (1999) CDK inhibitors: positive and negative regulators of G1-phase progression. Gen. Dev. **13**, 1501–1512
8. Rossow, P., Riddle, W. and Pardee, A. (1979) Synthesis of labile, serum-dependent protein in early G1 controls animal cell growth. Proc. Natl. Acad. Sci., U.S.A. **76**, 4446–4450
9. Campisi, J. amd Pardee, A.B. (1984) Post-transcriptional control of the onset of DNA synthesis by an insulin-like growth factor. Mol. Cell. Biol. **4**, 1807–1814
10. Besson, A., Dowdy, S. and Roberts, J. (2008) CDK inhibitors: cell cycle regulators and beyond. Dev. Cell **14**, 159–169
11. Rudolph, J. (2007) Cdc25 phosphatases: structure, specificity, and mechanism. Biochemistry **46**, 3595–3606
12. Iliakis, G., Wang, Y., Guan, J. and Wang, H. (2003) DNA damage checkpoint control in cells exposed to ionizing radiation. Oncogene **22**, 5834–5847
13. Stimac, E. and Morris, D.R. (1987) Messenger RNAs coding for enzymes of polyamine biosynthesis are induced during the G0-G1 transition, but not during traverse of the normal G1 phase. J. Cell. Phys. **133**, 590–594
14. Cress, A. and Gerner, E. (1980) Ornithine decarboxylase induction in cells stimulated to proliferate differs from that in continuously dividing cells. Biochem. J. **188**, 375–380
15. Katz, A. and Kahana, C. (1987) Activation of mammalian ornithine decarboxylase during stimulated growth. Mol. Cell. Biol. **7**, 2641–2643
16. Desiderio, M.A., Mattei, S., Biondi, G. and Colombo, M.P. (1993) Cytosolic and nuclear spermidine acetyltransferases in growing NIH 3T3 fibroblasts stimulated with serum or polyamines: relationship to polyamine-biosynthetic decarboxylases and histone acetyltransferase. Biochem. J. **293**, 475–479
17. Fredlund, J.O., Johansson, M.C., Dahlberg, E. and Oredsson, S.M. (1995) Ornithine decarboxylase and S-adenosylmethionine decarboxylase expression during the cell cycle of Chinese hamster ovary cells. Exp. Cell Res. **216**, 86–92
18. Hogan, B.L.M. (1971) Effect of growth conditions on the ornithine decarboxylase activity of rat hepatoma cells. Biochem. Biophys. Res. Commun. **45**, 301
19. Facchini, L.M. and Penn, L.Z. (1998) The molecular role of Myc in growth and transformation: recent discoveries lead to new insights. FASEB J. **12**, 633–651
20. Celano, P., Baylin, S.B. and Casero, Jr, R.A. (1989) Polyamines differentially modulate the transcription of growth-associated genes in human colon carcinoma cells. J. Biol. Chem. **264**, 8922–8927
21. Fuller, D., Gerner, E. and Russell, D. (1977) Polyamine biosynthesis and accumulation during the G1 to S phase transition. J. Cell. Phys. **93**, 81–88
22. Heby, O., Gray, J.W., Lindl, P.A., Marton, L.J. and Wilson, C.B. (1976) Changes in L-ornithine decarboxylase activity during the cell cycle. Biochem. Biophys. Res. Commun. **71**, 99–105

23. Hogan, B.L.M., Murden, S. and Blackledge, A. (1973) The effect of growth conditions on the synthesis and degradation of ornithine decarboxylase in cultured hepatoma cells. In Polyamines in Normal and Neoplastic Growth (Russell, D., ed.), pp. 239–246, Raven Press, New York
24. Kameji, T., Hayashi, S., Hoshino, K., Kakinuma, Y. and Igarashi, K. (1993) Multiple regulation of ornithine decarboxylase in enzyme-overproducing cells. Biochem. J. **289**, 581–586
25. Murakami, Y., Matsufuji, S., Kameji, T., Hayashi, S., Igarashi, K., Tamura, T., Tanaka, K. and Ichihara, A. (1992) Ornithine decarboxylase is degraded by the 26S proteasome without ubiquitination. Nature **360**, 597–599
26. Murakami, Y. and Hayashi, S. (1985) Role of antizyme in degradation of ornithine decarboxylase in HTC cells. Biochem. J. **226**, 893–896
27. Linden, M., Anehus, S., Långström, E., Baldetorp, B. and Heby, O. (1985) Cell cycle phase-dependent induction of ornithine decarboxylase-antizyme. J. Cell. Phys. **125**, 273–276
28. Pyronnet, S., Pradayrol, L. and Sonenberg, N. (2000) A cell cycle-dependent internal ribosome entry site. Mol. Cell **5**, 607–616
29. Nishimura, K., Sakuma, A., Yamashita, T., Hirokawa, G., Imataka, H., Kashiwagi, K. and Igarashi, K. (2007) Minor contribution of an internal ribosome entry site in the 5′-UTR of ornithine decarboxylase mRNA on its translation. Biochem. Biophys. Res. Commun. **364**, 124–130
30. Bettuzzi, S., Davalli, P., Astancolle, S., Pinna, C., Roncaglia, R., Boraldi, F., Tiozzo, R., Sharrard, M. and Corti, A. (1999) Coordinate changes of polyamine metabolism regulatory proteins during the cell cycle of normal human dermal fibroblasts. FEBS Lett. **446**, 18–22
31. Gemperlová, L., Cvikrová, M., Fischerová, L., Binarová, P., Fischer, L. and Eder, J. (2009) Polyamine metabolism during the cell cycle of synchronized tobacco BY-2 cell line. Plant Phys. Biochem. **2**, 584–591
32. Seiler, N. (2003) Thirty years of polyamine-related approaches to cancer therapy. Retrospect and prospect. Part 1. Selective enzyme inhibitors. Curr. Drug Targets **4**, 537–564
33. Seiler, N. (2003) Thirty years of polyamine-related approaches to cancer therapy. Retrospect and prospect. Part 2. Structural analogues and derivatives. Curr. Drug Targets **4**, 565–585
34. Casero, R.A. and Marton, L.J. (2007) Targeting polyamine metabolism and function in cancer and other hyperproliferative diseases. Nat. Rev. **6**, 373–390
35. Schaefer, E. and Seidenfeld, J. (1987) Effects of polyamine depletion on serum stimulation of quiescent 3T3 murine fibroblast cells. J. Cell. Phys. **133**, 546–552
36. Metcalf, B.W., Bey, P., Danzin, C., Jung, M.J., Casara, P. and Vevert, J.P. (1978) Catalytic irreversible inhibition of mammalian ornithine decarboxylase EC-41117 by substrate and product analogs. J. Am. Chem. Soc. **100**, 2551–2553
37. Fredlund, J.O. and Oredsson, S.M. (1996) Normal G1/S transition and prolonged S phase within one cell cycle after seeding cells in the presence of an ornithine decarboxylase inhibitor. Cell Prolif. **29**, 457–466
38. Regenass, U., Mett, H., Stanek, J., Mueller, M., Kramer, D. and Porter, C.W. (1994) CGP 48664, a new S-adenosylmethionine decarboxylase inhibitor with broad spectrum antiproliferative and antitumor activity. Cancer Res. **54**, 3210–3217
39. Fredlund, J.O. and Oredsson, S.M. (1997) Ordered cell cycle phase perturbations in Chinese hamster ovary cells treated with an S-adenosylmethionine decarboxylase inhibitor. Eur. J. Biochem. **249**, 232–238
40. Porter, C.W. and Bergeron, R.J. (1988) Enzyme regulation as an approach to interference with polyamine biosynthesis: an alternative to enzyme inhibition. Adv. Enzyme Reg. **27**, 57–79
41. Alm, K., Berntsson, P.S.H., Kramer, D.L., Porter, C.W. and Oredsson, S.M. (2000) Treatment of cells with the polyamine analog N^1,N^{11}-diethylnorspermine retards S phase progression within one cell cycle. Eur. J. Biochem. **267**, 4157–4164
42. Heby, O. (1981) Role of polyamines in the control of cell proliferation and differentiation. Differentiation **19**, 1–20
43. Laitinen, J., Stenius, K., Eloranta, T. and Hälttä, E. (1989) Polyamines may regulate S-phase progression but not the dynamic changes of chromatin during the cell cycle. J. Cell. Biochem. **68**, 200–212

44. Feuerstein, B.G., Pattabiraman, N. and Marton, L.J. (1990) Molecular mechanics of the interactions of spermine with DNA: DNA bending as a result of ligand binding. Nucleic Acids Res. **18**, 1271–1282
45. Grallert, B. and Boye, E. (2008) The multiple facets of the intra-S checkpoint. Cell Cycle **7**, 2315–2320
46. Johansson, V., Oredsson, S. and Alm, K. (2008) Polyamine depletion with two different polyamine analogues causes DNA damage in human breast cancer cell lines. DNA Cell Biol. **27**, 511–516
47. Liang, J. and Slingerland, J. (2003) Multiple roles of the PI3K/PKB (Akt) pathway in cell cycle progression. Cell Cycle **2**, 339–345
48. Koomoa, D., Yco, L., Borsics, T., Wallick, C. and Bachmann, A. (2008) Ornithine decarboxylase inhibition by α-difluoromethylornithine activates opposing signaling pathways via phosphorylation of both Akt/protein kinase B and p27Kip1 in neuroblastoma. Cancer Res. **68**, 9825–9831
49. Stanic, I., Facchini, A., Borzi, R.M., Stefanelli, C. and Flamigni, F. (2009) The polyamine analogue N^1,N^{11}-diethylnorspermine can induce chondrocyte apoptosis independently of its ability to alter metabolism and levels of natural polyamines. J. Cell. Phys. **219**, 109–116
50. Myhre, L., Alm, K., Hegardt, C., Staff, J., Jönsson, G., Larsson, S. and Oredsson, S.M. (2008) Different cell cycle kinetic effects of N^1,N^{11}-diethylnorspermine-induced polyamine depletion in four human breast cancer cell line. Anti-Cancer Drugs **19**, 359–368
51. Myhre, L., Alm, K., Johansson, M. and Oredsson, S.M. (2009) Reversible cell cycle kinetic effects of N^1,N^{11}-diethylnorspermine in the normal-like human breast epithelial cell line MCF-10A. Anti-Cancer Drugs **20**, 230–237
52. Kramer, D.L., Chang, B.D., Chen, Y., Diegelman, P., Alm, K., Black, A.R., Roninson, I.B. and Porter, C.W. (2001) Polyamine depletion in human melanoma cells leads to G1 arrest associated with induction of p21WAF1/CIP1/SDI1, changes in the expression of p21-regulated genes, and a senescence-like phenotype. Cancer Res. **61**, 7754–7762
53. Kramer, D.L., Vujcic, S., Diegelman, P., Alderfer, J., Miller, J.T., Black, J.D., Bergeron, R.J. and Porter, C.W. (1999) Polyamine analogue induction of the p53-p21WAF1/CIP1-Rb pathway and G1 arrest in human melanoma cells. Cancer Res. **59**, 1278–1286
54. Ray, R.M., Zimmerman, B.J., McCormack, S.A., Patel, T.B. and Johnson, L.R. (1999) Polyamine depletion alters the relationship of F-actin, G-actin, and thymosin β4 in migrating IEC-6 cells. Am. J. Phys. **276**, C684–C691
55. Chen, Y., Alm, K., Vujcic, S., Kramer, D.L., Kee, K., Diegelman, P. and Porter, C.W. (2003) The role of mitogen-activated protein kinase activation in determining cellular outcomes in polyamine analogue-treated human melanoma cells. Cancer Res. **63**, 3619–3625

Essays Biochem. (2009) **46**, 77–94; doi:10.1042/BSE0460006

6

Design of polyamine-based therapeutic agents: new targets and new directions

M.D. Thulani Senanayake*, Hemali Amunugama*, Tracey D. Boncher†, Robert A. Casero, Jr‡ and Patrick M. Woster*†[1]

**Department of Pharmaceutical Sciences, Eugene Applebaum College of Pharmacy and Health Sciences, Wayne State University, 259 Mack Ave, Detroit, MI 48202, U.S.A., †Ferris State University College of Pharmacy, PHR 105, 220 Ferris Drive, Big Rapids, MI 49307, U.S.A., and ‡The Sidney Kimmel Comprehensive Cancer Center at Johns Hopkins, The Johns Hopkins University School of Medicine, Bunting–Blaustein Cancer Research Building, 1650 Orleans Street, Baltimore, MD 21231, U.S.A.*

Abstract

Enzymes in the biosynthetic and catabolic polyamine pathway have long been considered targets for drug development, and early drug discovery efforts in the polyamine area focused on the design and development of specific inhibitors of the biosynthetic pathway, or polyamine analogues that specifically bind DNA. More recently, it has become clear that the natural polyamines are involved in numerous known and unknown cellular processes, and disruption of polyamine functions at their effector sites can potentially produce beneficial therapeutic effects. As new targets for polyamine drug discovery continue to evolve, the rational design of polyamine analogues will

[1]*To whom correspondence should be addressed (email pwoster@wayne.edu).*

result in more structurally diverse agents. In addition, the physical linkage of polyamine-like structures to putative drug molecules can have beneficial effects resulting from increases in DNA affinity and selective cellular uptake. The present chapter will summarize recent advances in the development of alkylpolyamine analogues as antitumour agents, and describe subsequent advances that have resulted from incorporating polyamine character into more diverse drug molecules. Specifically, new polyamine analogues, and the role of polyamine fragments in the design of antiparasitic agents, antitumour metal complexes, histone deacetylase inhibitors and lysine-specific demethylase 1 inhibitors, will be described.

Introduction: biosynthesis inhibitors and alkylpolyamine analogues

The polyamine pathway is an important target for drug design, since alteration of cellular polyamine levels results in the disruption of a variety of cellular functions [1]. Inhibitors of the polyamine pathway have traditionally been developed as potential antitumour and/or antiparasitic agents [1,2]. Specific inhibitors of polyamine biosynthesis have been useful as research tools to elucidate the cellular functions of the naturally occurring polyamines, but their success as therapeutic agents has been limited. This failure has been, in part, due to the ability of mammalian cells to compensate for inhibition of a single enzyme in the pathway by up-regulating other enzymes or by modulating polyamine transport. Specific inhibitors have now been developed for the enzymes in the biosynthetic pathway, ODC (ornithine decarboxylase), AdoMetDC [AdoMet (*S*-adenosylmethionine) decarboxylase] and the aminopropyltransferases spermidine synthase and spermine synthase. The structures of these classical polyamine biosynthesis inhibitors are shown in Figure 1. The compound DFMO (α-difluoromethylornithine; **1**), also known as eflornithine, is a mechanism-based inhibitor of ODC that was originally synthesized as an antitumour agent [3,4]. Clinical trials for DFMO as an antitumour agent were discontinued owing to lack of efficacy. However, there are some indications that it can be used for the treatment of glioblastoma [5,6] and prostate cancer [7], and it has gained recent attention as a chemopreventative agent [8,9]. DFMO is also approved for use in the treatment of West African trypanosomiasis caused by *Trypanosoma brucei gambiense*, but is ineffective against infections caused by *Trypanosoma brucei rhodesiense* (East African trypanosomiasis) [10–13]. It is currently marketed as a depilatory agent in the US. A number of effective inhibitors of AdoMetDC have been developed, but none have been marketed. The antileukaemic MGBG [methylglyoxal bis(guanylhydrazone); **2**], is a potent competitive inhibitor of mammalian AdoMetDC, with a K_i value of less than 1 μM [14]. However, MGBG is of limited use as a chemotherapeutic agent owing to a wide variety of other effects on cells, including induction of severe mitochondrial damage. More recently, conformationally restricted MGBG analogues such as the Ciba–Geigy

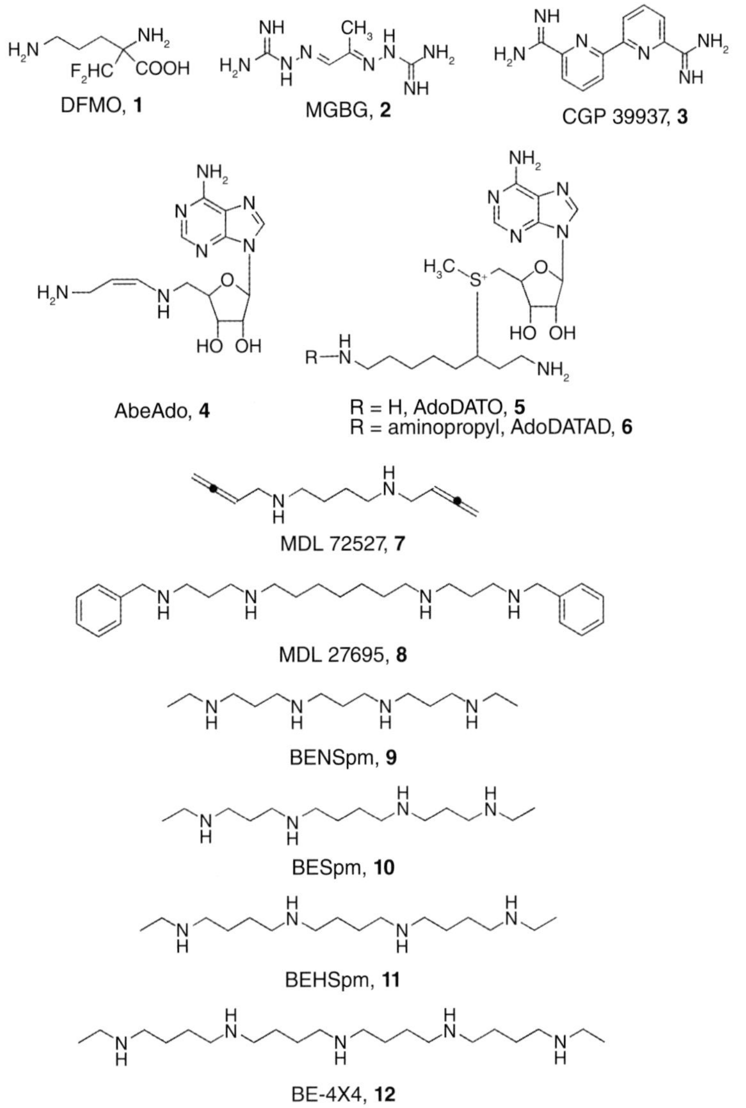

Figure 1. Structures of the classical inhibitors of polyamine biosynthesis

compound CGP 39937 (**3**) have been shown to have similar antitumour effects, but with reduced toxicity. One of the analogues in this series, CGP 48664, has advanced to Phase I and II human clinical trials. A number of structural analogues of AdoMet have also been synthesized. The most promising of these, MDL-73,811 or AbeAdo (**4**) is a potent enzyme-activated inhibitor of AdoMetDC [15]. AbeAdo has shown promise as an antitrypanosomal agent [16], but has not been developed for clinical use. Potent and specific inhibitors for the aminopropyltransferases spermidine synthase (AdoDATO; **5**) [17] and spermine synthase (AdoDATAD; **6**) [18,19] have been synthesized, but their pharmacokinetic properties precluded their development as drugs. A number of amine analogues have been reported to inhibit the polyamine oxidases SMO (spermine oxidase) and APAO (N^1-acetylpolyamine oxidase), including the well-established inhibitor MDL 72527 (**7**) [20]. To date, inhibitors specific for one of the two polyamine oxidases have not been identified.

In addition to enzyme inhibitors, a series of (bis)ethylpolyamines have been extensively studied as antitumour agents. Design of these analogues was based on the finding that natural polyamines utilize several feedback mechanisms which autoregulate their synthesis [1], and that they can be taken into cells by the polyamine transport system [21]. These analogues specifically down-regulate the synthesis of polyamines, but cannot substitute for the natural polyamines in their cell growth and survival functions [1,21]. The analogue MDL 27695 (**8**, Figure 1) is an early example of this type of analogue and possesses both antitumour and antiparasitic activity [22,23]. Among the most successful of these analogues (Figure 1) are BENSpm [bis(ethyl)norspermine; **9**], BESpm [bis(ethyl)spermine; **10**], BEHSpm [bis(ethyl)homospermine; **11**] and BE-4×4 [1,20-(ethylamino)-5,10,15-triazanonadecane; **12**]. The *N*,*N′*-bis(ethyl)polyamines are readily transported into mammalian cells [24], where they deplete cellular polyamines, decrease ODC and AdoMetDC activity, and can ultimately produce cytotoxicity, depending on the cell lines used [1,25,26]. These compounds have been shown to possess a wide variety of therapeutic effects that vary widely with surprisingly small structural changes [1]. More recent discovery efforts involving bis(ethyl)polyamines have resulted in a series of analogues with conformationally restricted central chains, and these analogues possess significant antitumour and antiparasitic activity [27–32]. Along similar lines, a series of bis(ethyl)oligamine analogues have been described that show promise as chemotherapeutic agents [33–35]. The first examples of unsymmetrically substituted alkylpolyamines were described in 1993 [36], and subsequent studies reveal that agents typified by CPENSpm (**13**), CHENSpm (**14**) and IPENSpm (**15**) (Figure 2) are also potent antitumour agents [1,37–42]. The library of terminally alkylated polyamine analogues has been extended to include more than 120 alkylpolyamines designed with varying polyamine backbone structures (3-4, 3-3-3, 3-4-4 and 3-7-3), and variation of the terminal alkyl substituents has been attempted to determine the optimal overall structure for antitumour effects. These agents produce a variety of cellular effects, and possess antitumour and/or antiparasitic activity *in vitro* [36,39–45]. Compounds **16–24** are representative of more than 25 unsymmetrically substituted alkylpolyamine analogues with 96 h IC_{50} values of less than 4 μM against the H157 non-small-cell lung tumour line. These analogues contain more structural diversity than previously seen, including aromatic moieties, unsaturations, stereochemistry and heteroatoms, indicating that exploration of the chemical space surrounding the terminal substituents will yield additional promising antitumour agents. Compounds **22**, **23** and **24** (also known as PN11400, PN11401 and PN11402 respectively), are currently in development as antitumour agents.

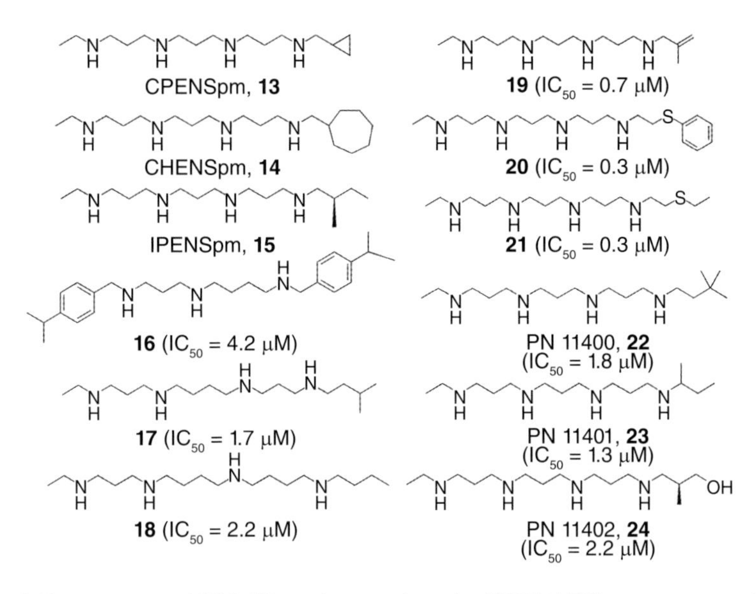

Figure 2. Structures and 96 h IC_{50} values against the NCI H157 lung tumour cell line for selected alkylpolyamines with antitumour activity

Polyamine-based antiparasitic agents

A number of alkylpolyamine analogues have been shown to possess impressive antiparasitic activity *in vitro*, and in some cases *in vivo* (Figure 3). Compounds **25** and **26** inhibited the growth of *Trypanosoma brucei brucei* (Lab110 EATRO strain) with 48 h IC_{50} values of 0.061 μM and 1.6 μM respectively, and maintained nearly identical submicromolar activity against the KETRI 243 and arsenic resistant KETRI 234 AS-10-3 clinical isolates of *T. brucei rhodesiense*. The analogue **27**, known as BW-1, also inhibited growth of the Lab 110 EATRO strain (48 h IC_{50}=0.24 μM), with equivalent activity against KETRI 243 (0.19 μM), KETRI 269 (0.75 μM) and KETRI 243 As-10-3 (0.20 μM) [1]. Compound **27** also inhibited the growth of the microsporidian *Enterocytozoon cuniculi* with a 48 h IC_{50} of 0.47 μM, and was curative in a mouse model for microsporidiosis [46]. It has also been shown that **27** is a competitive inhibitor and substrate for the microsporidial form of polyamine oxidase, and that, like the parent compound **8** [1], may act through activation by oxidation within the parasite [47]. Compound **28** is also an effective trypanocide, with a 48 h IC_{50} of 0.031 μM against Lab 110 EATRO, and 0.04 and 0.165 μM against KETRI 243 and KETRI 243 As-10-3 respectively. Interestingly, the 48 h IC_{50} value for the related analogue **29** (0.31 μM) is 10-fold higher that **28**, suggesting that bis substitution is optimal for antiparasitic activity *in vitro*. Finally, guanidines and biguanides such as compounds **30–32** have potent antitrypanosomal activity (Lab 110 EATRO 48 h IC_{50}=0.18, 0.09 and 0.18 μM respectively). These analogues are potent inhibitors of the parasitic enzyme trypanothione reductase, but have no activity against the human form of glutathione

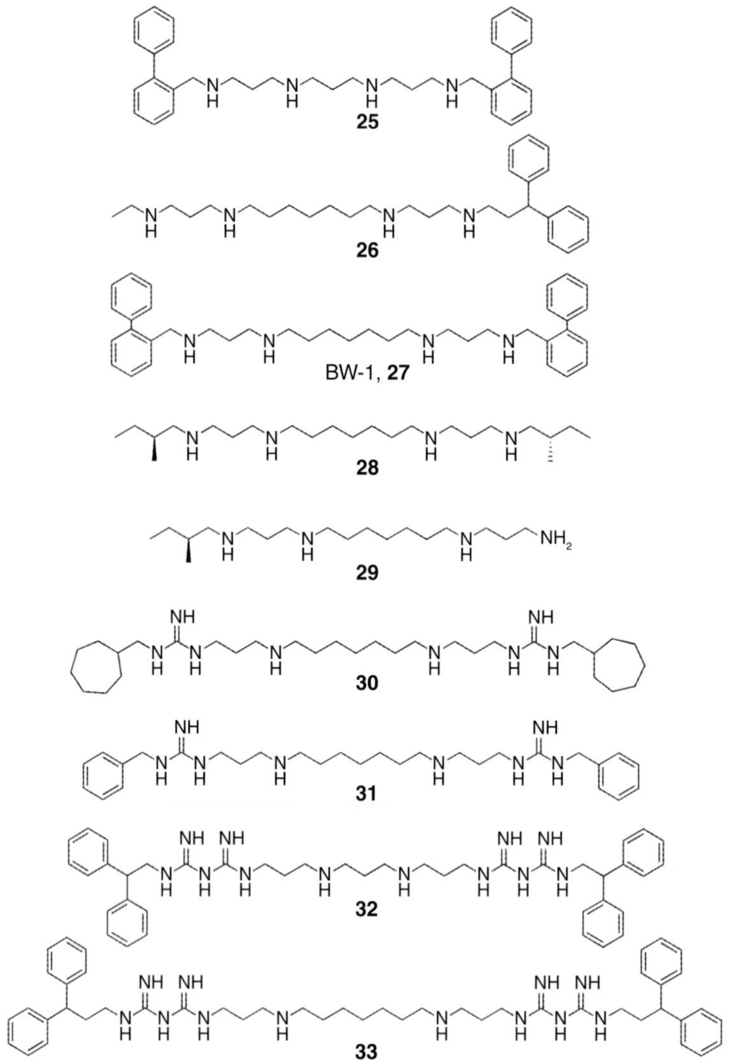

Figure 3. Structures of alkylpolyamine analogues with antiparasitic activity *in vitro* and/or *in vivo*.

reductase [44]. The related analogue **33** (also known as **2d**) inhibited Lab110 EATRO growth (48 h IC_{50}=0.62 μM), and has recently been shown to inhibit the growth of *Leishmania donovani* (48 h IC_{50}=6 μM, R. Madhubala and P. Woster, unpublished work).

Polyamine–metal complexes

Incorporation of polyamine side-chain residues into the structure of known antitumour agents that target DNA can lead to increases in antitumour activity. Dinuclear bis(platinum) complexes in which the metal centres were separated by diaminoalkanes are more potent than cisplatin against murine and human tumour cells *in vitro*, including cisplatin-resistant cell lines [48]. The trinuclear Pt (platinum) compound known as BBR 3464 (**34**) (Figure 4), in which three Pt centres are separated by diaminohexyl spacers [49,50] was shown

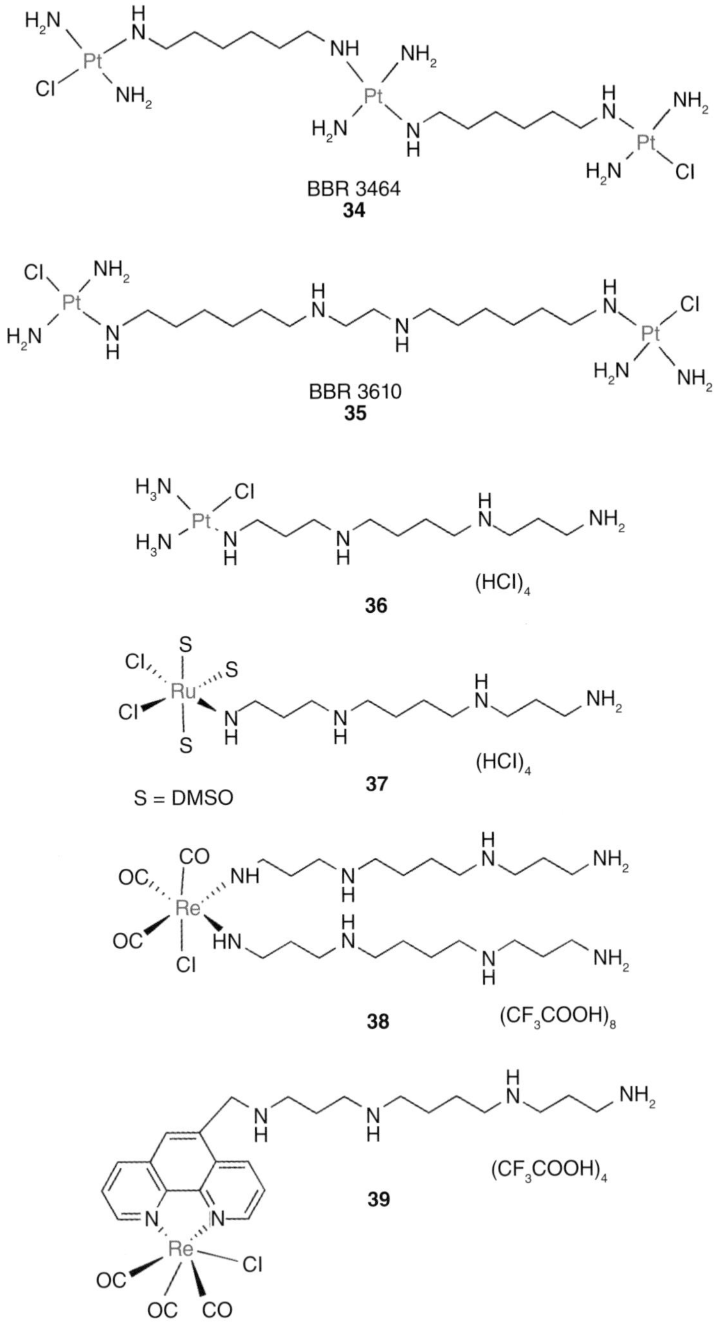

Figure 4. Polyamine–transition metal complexes with antitumour activity

to produce cytotoxicity in L1210 mouse leukaemia cells that was 30 times greater than cisplatin *in vitro* and to bind to DNA in a time-dependent manner [49] that was distinct from the binding pattern of cisplatin [50]. Recently, compound **35** (BBR3610) was shown to produce cytotoxicity superior to **34** through a caspase 8-dependent mechanism [51]. To date, the structure–activity relationships of polyamine–Pt complexes remain unexplored, and polyamine complexes with other transition metals, such as Re (rhenium), Rh (rhodium) and Ru (ruthenium), have not been synthesized and evaluated for antitumour activity. We postulated that transition metals complexed to polyamines with known affinity for DNA and established antitumour effects may be of value in the treatment of breast cancer, and other tumour types where cisplatin produces a poor therapeutic response. From an initial library of 25 polyamine–metal complexes containing Pt, Re, Rh or Ru centres, 11 analogues exhibited 96 h IC_{50} values of <10.0 μM in an H157 non-small-cell lung cancer MTT [3-(4,5-dimethylthiazol-2-yl)-2,5-diphenyl-2*H*-tetrazolium bromide] assay, with three of the 11 having 96 h IC_{50} values of <1.0 μM. Four analogues, Pt compound **36**, Ru compound **37** and Re compounds **38** and **39** (Figure 4), were selected for dose–response studies using an MTT assay in the MCF7 breast tumour cell line, in which they exhibited 96 h IC_{50} values of 6.6, 3.8, 9.3 and 6.9 μM respectively. By contrast, the 96 h IC_{50} value for cisplatin was >250 μM under the same conditions. These results demonstrate that polyamine complexes containing three distinct (Pt, Ru and Re) metal centres possess antitumour activity far superior to cisplatin against MCF7 breast tumour cells *in vitro*. Polyamine–metal complexes **36–39** formed more extensive cross-links with plasmid DNA than cisplatin in a time- (6 h) and concentration- (3.0–300 μM) dependent manner. These results clearly demonstrate that polyamine–metal complexes are more potent than the cisplatin in breast tumour cells *in vitro*, and show a greater ability to bind to DNA.

Polyamine-based inhibitors of HDAC (histone deacetylase)

It is widely known that normal mammalian cells exhibit an exquisite level of control of chromatin architecture by maintaining a balance between HAT (histone acetyltransferase) and HDAC activity. In some tumour cells, hypoacetylation of histones results in aberrant gene silencing leading to the underexpression of growth regulatory proteins, and contributing to the development of cancer. HDAC inhibitors such as TSA (trichostatin A), MS-275 {*N*-(2- aminophenyl) -4-[*N*-(pyridin-3-ylmethoxycarbonyl)-aminomethyl]benzamide} and SAHA (suberoylanilide hydroxamic acid) can cause growth arrest in a wide range of transformed cells, and can inhibit the growth of human tumour xenografts [52]. Clinical studies indicate that HDAC inhibitors, such as SAHA and MS-275, are effective therapies for human cancer, but dose-limiting toxicity from the HDAC inhibitor remains a problem [53]. Traditional class I, IIa, IIb and IV HDAC inhibitors possess three structural features that are required for optimal activity: an aromatic cap group, an aliphatic chain and a metal-binding functional group (Figure 5). We hypothesized that addition of a polyamine side chain would increase the affinity of the agent for chromatin and facilitate import into cells via the polyamine transport system. We also felt that it would be possible to vary the structure of the terminal substituent on the polyamine side chain so as to target each of the 11 class I, IIa, IIb and IV HDAC isoforms [52] specifically. An initial library of 16 PAHA (polyaminohydroxamic acid) analogues was thus synthesized and screened for inhibitory activity against a global mixture of HDACs derived from HeLa cell lysates [54]. Compounds **40** and **41** (Figure 5) inhibited HDAC in this assay by 75% or greater at 1.0 μM. Compared with MS-275, **41** produced significantly greater induction of acetylated histone H3 and H4, and promoted greater re-expression of the cyclin-dependent kinase inhibitor p21^{Waf1}, in the ML-1 mouse leukaemia

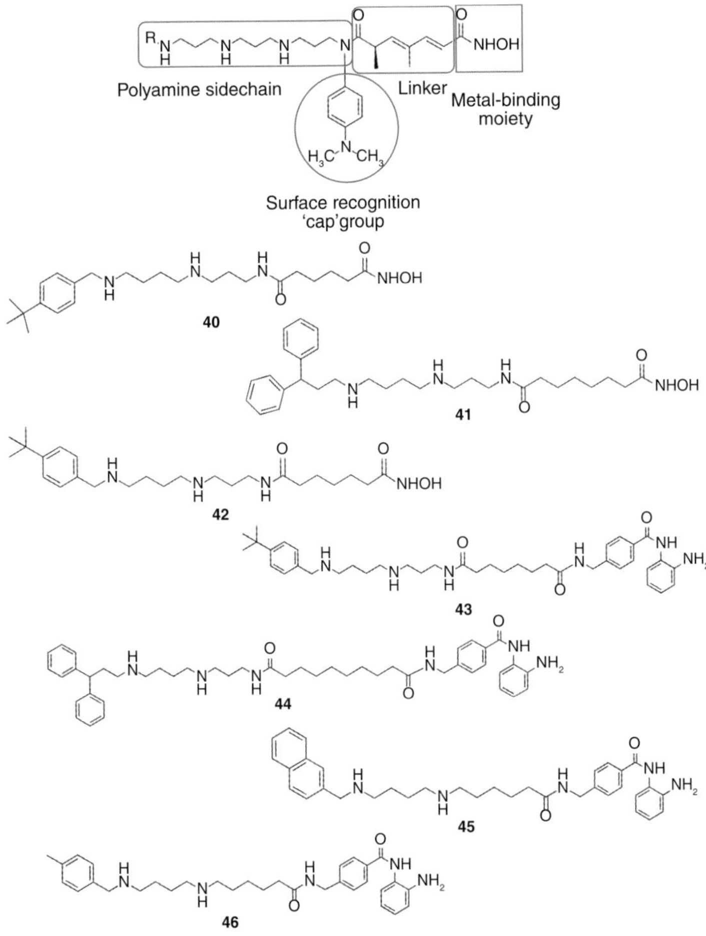

Figure 5. General structure for polyamine-based HDAC inhibitors, and selected PAHA and PABA analogues

cell line. In addition, 1.0 μM **42** inhibited the HDAC mixture by 51.5%, but caused a 253-fold induction of acetylated α-tubulin, accompanied by minimal increases in acetylated histones H3, H4 and $p21^{Waf1}$ [55]. These results strongly suggest that **42** shows a marked selectivity towards HDAC6, which is known to deacetylate α-tubulin. Although **40–42** obviously were able to enter mammalian cells, they did not serve as substrates for the polyamine transport system, presumably because of the negative charge of the hydroxamic acid moiety under the assay conditions.

To explore additional metal-binding moieties that could facilitate cellular transport, a series of 40 PABAs (polyaminobenzamides) and their homologues were synthesized and evaluated as inhibitors of global HDAC [55]. Representative structures for this class are shown in Figure 5. Compound **43** exhibited an IC_{50} value against global HDAC of 4.9 μM, which is comparable with the reported IC_{50} value for MS-275 (4.8 μM). Analogues in the PABA series were evaluated against four HDAC isoforms representing Class I (HDAC 1, 3 and 8) and Class II (HDAC6). In these analogues, structural modifications were made in the linker chain length, in the polyamine substituent and in some cases in the metal-binding moiety. Isoform selectivity among the four HDACs evaluated varied significantly, demonstrating that the global percentage of HDAC inhibition is a composite of strong and weak inhibition at different isoforms and suggesting that the observed HDAC selectivity with PAHAs such as **42**, and PABAs such as **43** may be in part due to structural variations in their polyamine side chains. As mentioned above, one of the potential advantages of incorporating polyamine side chains into PAHA and PABA HDAC inhibitors was that the resulting molecules could utilize the polyamine transport system. Our preliminary data demonstrates that PABAs such as **43** are effectively imported using the polyamine transport system, as verified by [^{14}C]spermidine uptake competition assays [55].

Three of the most active global HDAC inhibitors in the PABA series [52], compounds **44**, **45** and **46**, were markedly selective for HDAC1, and were evaluated against MCF7 wild-type tumour cells *in vitro*. Over a range of concentrations between 0.3 and 30 mM, PABA **44** was inactive, whereas **45** was cytostatic. However, PABA **46** was cytotoxic in the MCF7 cell line, with a 96 h IC_{50} of 0.9 μM. In the MCF10A non-tumorigenic breast epithelial cell line, **46** exhibited a 96 h IC_{50} of 24 μM, thus demonstrating selectivity for tumour cells. Under the same conditions, SAHA exhibited 96 h IC_{50} values of 8.5 mM (MCF7) and 31 mM (MCF10A), and thus **46** compares favourably with known HDAC inhibitors that are currently in use in the clinic. Importantly, **44–46** were efficiently imported by the polyamine transporter, as determined by [^{14}C]spermidine-uptake competition assays. Microscopic examination of treated cells revealed that cytotoxicity was mediated by apoptosis in MCF7 cells treated with **46**. Subsequent experiments showed that **46**, but not **44** or **45**, promoted the induction of ANXA1 (annexin A1). HDAC inhibitors have been shown to promote apoptosis through induction of ANXA1

[56], and recent studies suggest that breast tumours expressing high levels of ANXA1 are more likely to respond to chemotherapy than cells with low levels of ANXA1 [57]. Additional research is required to determine whether there is a functional relationship between inhibition of a specific HDAC isoform and induction of ANXA1.

Polyamine-based inhibitors of LSD1 (lysine-specific demethylase 1)

The potential role of polyamine analogues as inhibitors of LSD1, as well as the epigenetic effects such analogues may have on the re-expression of aberrantly silenced genes, has been described in detail elsewhere in this *Essays in Biochemistry* volume (chapter 7 by Casero). LSD1 was identified in part because its C-terminal domain shares significant sequence homology with the polyamine oxidases APAO and SMO [58,59]. Several groups have identified polyamine analogues that act as inhibitors of these two polyamine oxidases. MGBG (**2**) (Figure 1), a classical inhibitor of polyamine metabolism, was a potent inducer of APAO, but not SMO, in rat liver [60]. MDL 72,527 (**7**) (Figure 1), is a potent inhibitor of murine polyamine oxidase [61], as well as human APAO and SMO, but does not inhibit MAO [59,62]. It has previously been demonstrated that the polyaminoguanidine guazatine is a non-competitive inhibitor of maize polyamine oxidase [63]. Taken together, these results suggest that potent and selective inhibitors for the homologous flavin-dependent amine oxidase LSD1 can also be designed and synthesized that contain the structural motifs described above.

We have reported the synthesis of a novel series of polyamino (bis)guanidines and polyaminobiguanides [44] that are highly effective inhibitors of the parasitic enzyme trypanothione reductase, but that do not affect the human counterpart enzyme glutathione reductase. These compounds act as potent antitrypanosomal agents *in vitro*, with 48 h IC_{50} values against *T. brucei* as low as 90 nM. Based on the observation that polyaminoguanidines such as guazatine are potent inhibitors of amine oxidases, we also evaluated these compounds for the ability to inhibit the closely related enzyme LSD1. These studies established that polyaminoguanidines and polyaminobiguanides such as **47** and **33** respectively (Figure 6), act as non-competitive inhibitors of LSD1 [64]. Based on the inhibitory activity of **47**, five additional guanidine analogues, **48–52** (Figure 6), were synthesized and evaluated for the ability to increase H3K4me2 (histone 3 dimethyl-lysine 4), a direct indicator of LSD1 inhibition. After a 48-h exposure, 5 μM compound **49** produced a 2.3-fold increase in the level of H3K4me2 in the KG1a haematopoietic cell line, but not in HL60 human promyelocytic leukaemia cells. Compound **50** caused a 1.6-fold increase in H3K4me2 in the KG1a line, but only at 10 μM, and did not affect H3K4me2 levels in the HL60 line. Compounds **48**, **51** and **52** had no effect in either cell line. These results suggest that replacement of the four methyl substituents in **47** with more bulky groups leads to a drastic reduction

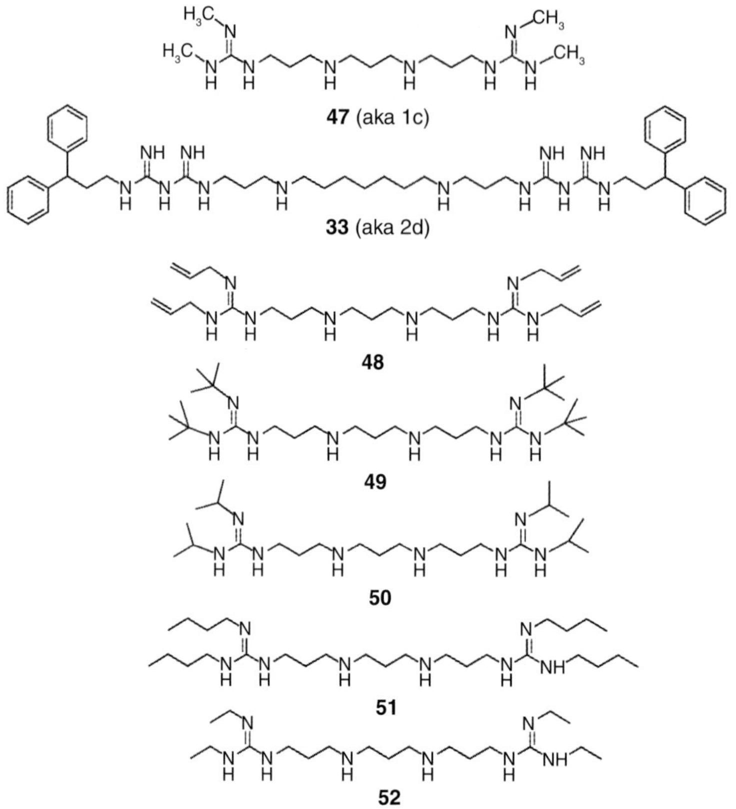

Figure 6. Structures of compound 47 (1c), 33 (2d) and 1c analogues 48–52

in LSD1 inhibition, and subsequently there is little effect on H3K4me2 levels. The synthesis of additional analogues in the polyaminoguanidine and polyaminobiguanide series is ongoing.

Conclusions and future directions

Prior to 1990, antitumour drug discovery research in the polyamine area was characterized by the design, synthesis and development of polyamine biosynthesis inhibitors. Despite a great deal of research, development of specific inhibitors as antitumour agents has only resulted in advancing one compound to the market. Drug discovery efforts in the polyamine field have begun to move away from the design of inhibitors for specific enzymes in the polyamine pathway, and to some degree away from the synthesis of analogues of the natural polyamines. However, the inclusion of polyamine-like structures in putative drug molecules allows them to take advantage of polyamine transport, and increases their affinity for chromatin. Thus agents with known therapeutic effects can be made more specific for target cells by virtue of mimicking the properties of the natural polyamines. Analogues of this type will capitalize on their polyamine character to target them to specific sites, but the activity of these derivatives will not depend on the polyamine structure within the molecule.

Terminally alkylated polyamine analogues continue to hold promise as therapies for a range of human cancers. A relatively large number of symmetrically substituted alkylpolyamines have been synthesized, and a number of these have been advanced to pre-clinical trials, and in some cases human clinical trials. In human clinical trials, some adverse effects have been noted, especially neurological symptoms (unilateral weakness, dysphagia, dysarthria, numbness, paresthesias, and ataxia [65], and aphasia, dizziness, vertigo and slurred speech [66]). Other Phase I studies in patients with non-small-cell lung cancer indicate that bis(ethyl)polyamines can be administered safely, and dose-limiting toxicities are mainly gastrointestinal [67]. Because the natural polyamines are ubiquitous in human cells, are strong cations and perform a variety of functions, it is not surprising that these analogues produce off-target effects. Through the use of biochemical and molecular biological techniques, combined with microarray and proteomics data, it will be possible to associate specific molecular targets with each cellular effect of the natural polyamines. As new effector sites for the natural polyamines are uncovered, a collaborative effort between chemists and biologists will facilitate optimization of analogue structure for each of these sites, thus reducing off-target effects and affording more specific therapeutic agents. Rational drug design principles can ultimately be used to identify a wide variety of novel therapeutic agents.

Summary

- *Classical polyamine analogues, including biosynthesis inhibitors and linear analogues of the natural polyamines, have been extensively studied as antitumour and/or antiparasitic agents. Despite this intensive level of study, only a single agent, DFMO, has advanced to the clinic. However, 'second generation' alkylpolyamines containing greater structural diversity show great promise as chemotherapeutic agents.*
- *Selected polyamine analogues with substitutions on the terminal nitrogens, such as BW-1, have excellent antiparasitic activity in multiple organisms. In addition, guanidine- and biguanide-based analogues that are structurally related to alkylpolyamines have potent antiparasitic activity.*
- *Polyamines complexed to the transition metal platinum have been developed that produce significant antineoplastic effects. More recently, polyamine complexes with platinum, and with other transition metals such as rhenium, rhodium or ruthenium, are more potent that cisplatin, and may be useful in the treatment of breast cancer.*

- *Incorporation of a polyamine side chain into the structure of known HDAC inhibitors produced analogues that are as potent or more potent than existing HDAC inhibitors. In addition, these analogues are targeted to chromatin, can utilize the polyamine transport system to enter cells, and in some cases show good selectivity for individual HDAC isoforms. The effects of these analogues may be mediated through induction of the pro-apoptotic protein factor ANXA1.*
- *Polyaminoguanidines and polyaminobiguanides previously identified as antitrypanosomal agents are structurally similar to guanidines that inhibit amine oxidases. As such, these analogues were evaluated as inhibitors of the recently discovered enzyme LSD1. These compounds produce significant epigenetic changes in tumour cells that lead to the re-expression of tumour suppressor factors that are important in human cancer.*

References

1. Casero, R.A. and Woster, P.M. (2001) Terminally alkylated polyamine analogues as chemotherapeutic agents. J. Med. Chem. **44**, 1–26
2. Gerner, E.W. and Meyskens, Jr, F.L. (2004) Polyamines and cancer: old molecules, new understanding. Nat. Rev. Cancer **4**, 781–792
3. Bey, P., Danzin, C., Van Dorsselaer, V., Mamont, P., Jung, M. and Tardif, C. (1978) Analogues of ornithine as inhibitors of ornithine decarboxylase. New deductions concerning the topography of the enzyme's active site. J. Med. Chem. **21**, 50–55
4. Mamont, P.S., Duchesne, M.C., Grove, J. and Bey, P. (1978) Anti-proliferative properties of DL-α-difluoromethyl ornithine in cultured cells. A consequence of the irreversible inhibition of ornithine decarboxylase. Biochem. Biophys. Res. Commun. **81**, 58–66
5. Levin, V.A., Hess, K.R., Choucair, A., Flynn, P.J., Jaeckle, K.A., Kyritsis, A.P., Yung, W.K., Prados, M.D., Bruner, J.M., Ictech, S. et al. (2003) Phase III randomized study of postradiotherapy chemotherapy with combination α-difluoromethylornithine-PCV versus PCV for anaplastic gliomas. Clin. Cancer Res. **9**, 981–990
6. Levin, V.A., Uhm, J.H., Jaeckle, K.A., Choucair, A., Flynn, P.J., Yung, W.K. A., Prados, M.D., Bruner, J.M., Chang, S.M., Kyritsis, A.P. et al. (2000) Phase III randomized study of postradiotherapy chemotherapy with α-difluoromethylornithine-procarbazine, *N*-(2-chloroethyl)-*N'*-cyclohexyl-*N*-nitrosurea, vincristine (DFMO-PCV) versus PCV for glioblastoma multiforme. Clin. Cancer Res. **6**, 3878–3884
7. Simoneau, A.R., Gerner, E.W., Nagle, R., Ziogas, A., Fujikawa-Brooks, S., Yerushalmi, H., Ahlering, T.E., Lieberman, R., McLaren, C.E., Anton-Culver, H. and Meyskens, Jr, F.L. (2008) The effect of difluoromethylornithine on decreasing prostate size and polyamines in men: results of a year-long phase IIb randomized placebo-controlled chemoprevention trial. Cancer Epidemiol. Biomarkers Prev. **17**, 292–299
8. Raul, F., Gosse, F., Osswald, A.B., Bouhadjar, M., Foltzer-Jourdainne, C., Marescaux, J. and Soler, L. (2007) Follow-up of tumor development in the colons of living rats and implications for chemoprevention trials: assessment of aspirin-difluoromethylornithine combination. Int. J. Oncol. **31**, 89–95
9. Meyskens, Jr, F.L., McLaren, C.E., Pelot, D., Fujikawa-Brooks, S., Carpenter, P.M., Hawk, E., Kelloff, G., Lawson, M.J., Kidao, J., McCracken, J. et al. (2008) Difluoromethylornithine plus sulindac for the prevention of sporadic colorectal adenomas: a randomized placebo-controlled, double-blind trial. Cancer Prev. Res. **1**, 32–38

10. Bacchi, C.J., Nathan, H.C., Hutner, S.H., McCann, P.P. and Sjoerdsma, A. (1980) Polyamine metabolism: a potential therapeutic target in trypanosomes. Science **210**, 332–334
11. Bacchi, C.J., Nathan, H.C., Livingston, T., Valladares, G., Saric, M., Sayer, P.D., Njogu, A.R. and Clarkson, Jr, A.B. (1990) Differential susceptibility to DL-α-difluoromethylornithine in clinical isolates of *Trypanosoma brucei rhodesiense*. Antimicrob. Agents Chemother. **34**, 1183–1188
12. Heby, O., Persson, L. and Rentala, M. (2007) Targeting the polyamine biosynthetic enzymes: a promising approach to therapy of African sleeping sickness, Chagas' disease, and leishmaniasis. Amino Acids. **33**, 359–366
13. Bacchi, C.J., Garofalo, J., Ciminelli, M., Rattendi, D., Goldberg, B., McCann, P.P. and Yarlett, N. (1993) Resistance to DL-α-difluoromethylornithine by clinical isolates of *Trypanosoma brucei rhodesiense*. Role of *S*-adenosylmethionine. Biochem. Pharmacol. **46**, 471–481
14. Janne, J., Alhonen, L. and Leinonen, P. (1991) Polyamines: from molecular biology to clinical applications. Ann. Med. **23**, 241–259
15. Danzin, C., Marchal, P. and Casara, P. (1990) Irreversible inhibition of rat *S*-adenosylmethionine decarboxylase by 5′-([(Z)-4-amino-2-butenyl]methylamino)-5′-deoxyadenosine. Biochem. Pharmacol. **40**, 1499–1503
16. Bacchi, C.J., Nathan, H.C., Yarlett, N., Goldberg, B., McCann, P.P., Bitonti, A.J. and Sjoerdsma, A. (1992) Cure of murine *Trypanosoma brucei rhodesiense* infections with an *S*-adenosylmethionine decarboxylase inhibitor. Antimicrob. Agents Chemother. **36**, 2736–2740
17. Tang, K.C., Mariuza, R. and Coward, J.K. (1981) Synthesis and evaluation of some stable multisubstrate adducts as specific inhibitors of spermidine synthase. J. Med. Chem. **24**, 1277–1284
18. Pegg, A.E., Wechter, R., Poulin, R., Woster, P.M. and Coward, J.K. (1989) Effect of *S*-adenosyl-1,12-diamino-3-thio-9-azadodecane, a multisubstrate adduct inhibitor of spermine synthase, on polyamine metabolism in mammalian cells. Biochemistry **28**, 8446–8453
19. Woster, P.M., Black, A.Y., Duff, K.J., Coward, J.K. and Pegg, A.E. (1989) Synthesis and biological evaluation of *S*-adenosyl-1,12-diamino-3-thio-9-azadodecane, a multisubstrate adduct inhibitor of spermine synthase. J. Med. Chem. **32**, 1300–1307
20. Bianchi, M., Polticelli, F., Ascenzi, P., Botta, M., Federico, R., Mariottini, P. and Cona, A. (2006) Inhibition of polyamine and spermine oxidases by polyamine analogues. FEBS J. **273**, 1115–1123
21. Porter, C.W. and Sufrin, J.R. (1986) Interference with polyamine biosynthesis and/or function by analogs of polyamines or methionine as a potential anticancer chemotherapeutic strategy. Anticancer Res. **6**, 525–542
22. Bitonti, A.J., Dumont, J.A., Bush, T.L., Edwards, M.L., Stemerick, D.M., McCann, P.P. and Sjoerdsma, A. (1989) Bis(benzyl)polyamine analogs inhibit the growth of chloroquine-resistant human malaria parasites (*Plasmodium falciparum*) *in vitro* and in combination with α-difluoromethylornithine cure murine malaria. Proc. Natl. Acad. Sci. U.S.A. **86**, 651–655
23. Edwards, M.L., Snyder, R.D. and Stemerick, D.M. (1991) Synthesis and DNA-binding properties of polyamine analogues. J. Med. Chem. **34**, 2414–2420
24. Porter, C.W., Berger, F.G., Pegg, A.E., Ganis, B. and Bergeron, R.J. (1987) Regulation of ornithine decarboxylase activity by spermidine and the spermidine analogue N1N8-bis(ethyl)spermidine. Biochem. J. **242**, 433–440
25. Bergeron, R.J., Neims, A.H., McManis, J.S., Hawthorne, T.R., Vinson, J.R., Bortell, R. and Ingeno, M.J. (1988) Synthetic polyamine analogues as antineoplastics. J. Med. Chem. **31**, 1183–1190
26. Porter, C.W., Pegg, A.E., Ganis, B., Madhabala, R. and Bergeron, R.J. (1990) Combined regulation of ornithine and *S*-adenosylmethionine decarboxylases by spermine and the spermine analogue *N*1 *N*12- bis(ethyl)spermine. Biochem. J. **268**, 207–212
27. Frydman, B., Blokhin, A.V., Brummel, S., Wilding, G., Maxuitenko, Y., Sarkar, A., Bhattacharya, S., Church, D., Reddy, V.K., Kink, J.A. et al. (2003) Cyclopropane-containing polyamine analogues are efficient growth inhibitors of a human prostate tumor xenograft in nude mice. J. Med. Chem. **46**, 4586–4600
28. Frydman, B., Porter, C.W., Maxuitenko, Y., Sarkar, A., Bhattacharya, S., Valasinas, A., Reddy, V.K., Kisiel, N., Marton, L.J. and Basu, H.S. (2003) A novel polyamine analog (SL-11093) inhibits growth of human prostate tumor xenografts in nude mice. Cancer Chemother. Pharmacol. **51**, 488–492

29. Hacker, A., Marton, L.J., Sobolewski, M. and Casero, Jr, R.A. (2008) *In vitro* and *in vivo* effects of the conformationally restricted polyamine analogue CGC-11047 on small cell and non-small cell lung cancer cells. Cancer Chemother. Pharmacol. **63**, 45–53
30. Reddy, V.K., Valasinas, A., Sarkar, A., Basu, H.S., Marton, L.J. and Frydman, B. (1998) Conformationally restricted analogues of 1*N*,12*N*-bisethylspermine: synthesis and growth inhibitory effects on human tumor cell lines. J. Med. Chem. **41**, 4723–4732
31. Valasinas, A., Sarkar, A., Reddy, V.K., Marton, L.J., Basu, H.S. and Frydman, B. (2001) Conformationally restricted analogues of 1*N*,14*N*-bisethylhomospermine (BE-4-4-4): synthesis and growth inhibitory effects on human prostate cancer cells. J. Med. Chem. **44**, 390–403
32. Waters, W.R., Frydman, B., Marton, L.J., Valasinas, A., Reddy, V.K., Harp, J.A., Wannemuehler, M.J. and Yarlett, N. (2000) [N^1,N^{12}]bis(ethyl)-cis-6,7-dehydrospermine: a new drug for treatment and prevention of *Cryptosporidium parvum* infection of mice deficient in T-cell receptor α. Antimicrob. Agents Chemother. **44**, 2891–2894
33. Valasinas, A., Reddy, V.K., Blokhin, A.V., Basu, H.S., Bhattacharya, S., Sarkar, A., Marton, L.J. and Frydman, B. (2003) Long-chain polyamines (oligoamines) exhibit strong cytotoxicities against human prostate cancer cells. Bioorg. Med. Chem. **11**, 4121–4131
34. Huang, Y., Hager, E.R., Phillips, D.L., Dunn, V.R., Hacker, A., Frydman, B., Kink, J.A., Valasinas, A.L., Reddy, V.K., Marton, L.J. et al. (2003) A novel polyamine analog inhibits growth and induces apoptosis in human breast cancer cells. Clin. Cancer Res. **9**, 2769–2777
35. Huang, Y., Keen, J.C., Hager, E., Smith, R., Hacker, A., Frydman, B., Valasinas, A.L., Reddy, V.K., Marton, L.J., Casero, Jr, R.A. and Davidson, N.E. (2004) Regulation of polyamine analogue cytotoxicity by c-Jun in human MDA-MB-435 cancer cells. Mol. Cancer Res. **2**, 81–88
36. Saab, N.H., West, E.E., Bieszk, N.C., Preuss, C.V., Mank, A.R., Casero, Jr, R.A. and Woster, P.M. (1993) Synthesis and evaluation of unsymmetrically substituted polyamine analogues as modulators of human spermidine/spermine-N^1- acetyltransferase (SSAT) and as potential antitumor agents. J. Med. Chem. **36**, 2998–3004
37. Ha, H.C., Woster, P.M. and Casero, Jr, R.A. (1998) Unsymmetrically substituted polyamine analogue induces caspase-independent programmed cell death in Bcl-2-overexpressing cells. Cancer Res. **58**, 2711–2714
38. Ha, H.C., Woster, P.M., Yager, J.D. and Casero, Jr, R.A. (1997) The role of polyamine catabolism in polyamine analogue-induced programmed cell death. Proc. Natl. Acad. Sci. U.S.A. **94**, 11557–11562
39. McCloskey, D.E., Casero, Jr, R.A. Woster, P.M. and Davidson, N.E. (1995) Induction of programmed cell death in human breast cancer cells by an unsymmetrically alkylated polyamine analogue. Cancer Res. **55**, 3233–3236
40. McCloskey, D.E., Woster, P.M., Casero, Jr, R.A. and Davidson, N.E. (2000) Effects of the polyamine analogues *N*1-ethyl-*N*11-[(cyclopropyl)methyl]- 4,8-diazaundecane and *N*1-ethyl-*N*11-[(cycloheptyl) methyl]-4,8- diazaundecane in human prostate cancer cells. Clin. Cancer Res. **6**, 17–23
41. McCloskey, D.E., Yang, J., Woster, P.M., Davidson, N.E. and Casero, Jr, R.A. (1996) Polyamine analogue induction of programmed cell death in human lung tumor cells. Clin. Cancer Res. **2**, 441–446
42. Webb, H.K., Wu, Z., Sirisoma, N., Ha, H.C., Casero, Jr, R.A. and Woster, P.M. (1999) 1-(*N*-alkylamino)-11-(*N*-ethylamino)-4,8-diazaundecanes: simple synthetic polyamine analogues that differentially alter tubulin polymerization. J. Med. Chem. **42**, 1415–1421
43. Bellevue, III, F.H., Boahbedason, M., Wu, R., Woster, P.M., Casero, Jr, R.A., Rattendi, D., Lane, S. and Bacchi, C.J. (1996) Structural comparison of alkylpolyamine analogues with potent *in vitro* antitumor or antiparasitic activity. Bioorg. Med. Chem. Lett. **6**, 2765–2770.
44. Bi, X., Lopez, C., Bacchi, C.J., Rattendi, D. and Woster, P.M. (2006) Novel alkylpolyaminoguanidines and alkylpolyaminobiguanides with potent antitrypanosomal activity. Bioorg. Med. Chem. Lett. **16**, 3229–3232
45. Davidson, N.E., Hahm, H.A., McCloskey, D.E., Woster, P.M. and Casero, Jr, R.A. (1999) Clinical aspects of cell death in breast cancer: the polyamine pathway as a new target for treatment. Endocr. Relat. Cancer **6**, 69–73

46. Zou, Y., Wu, Z., Sirisoma, N., Woster, P.M., Casero, Jr, R.A., Weiss, L.M., Rattendi, D., Lane, S. and Bacchi, C.J. (2001) Novel alkylpolyamine analogues that possess both antitrypanosomal and antimicrosporidial activity. Bioorg. Med. Chem. Lett. **11**, 1613–1617
47. Bacchi, C.J., Yarlett, N., Faciane, E., Bi, X., Rattendi, D., Weiss, L.M. and Woster, P.M. (2009) Metabolism of an alkyl polyamine analog by a polyamine oxidase from the microsporidian *Encephalitozoon cuniculi*. Antimicrob. Agents Chemother. **53**, 2599–2604
48. Kraker, A.J., Hoeschele, J.D., Elliott, W.L., Showalter, H.D., Sercel, A.D. and Farrell, N.P. (1992) Anticancer activity in murine and human tumor cell lines of bis(platinum) complexes incorporating straight-chain aliphatic diamine linker groups. J. Med. Chem. **35**, 4526–4532
49. Di Blasi, P., Bernareggi, A., Beggiolin, G., Piazzoni, L., Menta, E. and Formento, M.L. (1998) Cytotoxicity, cellular uptake and DNA binding of the novel trinuclear platinun complex BBR 3464 in sensitive and cisplatin resistant murine leukemia cells. Anticancer Res. **18**, 3113–3117
50. Brabec, V., Kasparkova, J., Vrana, O., Novakova, O., Cox, J.W., Qu, Y. and Farrell, N. (1999) DNA modifications by a novel bifunctional trinuclear platinum phase I anticancer agent. Biochemistry **38**, 6781–6790
51. Mitchell, C., Kabolizadeh, P., Ryan, J., Roberts, J.D., Yacoub, A., Curiel, D.T., Fisher, P.B., Hagan, M.P., Farrell, N.P., Grant, S. and Dent, P. (2007) Low-dose BBR3610 toxicity in colon cancer cells is p53-independent and enhanced by inhibition of epidermal growth factor receptor (ERBB1)-phosphatidyl inositol 3 kinase signaling. Mol. Pharmacol. **72**, 704–714
52. Johnstone, R.W. (2002) Histone-deacetylase inhibitors: novel drugs for the treatment of cancer. Nat. Rev. **1**, 287–299
53. Ryan, Q.C., Headlee, D., Acharya, M., Sparreboom, A., Trepel, J.B., Ye, J., Figg, W.D., Hwang, K., Chung, E.J., Murgo, A. et al. (2005) Phase I and pharmacokinetic study of MS-275, a histone deacetylase inhibitor, in patients with advanced and refractory solid tumors or lymphoma. J. Clin. Oncol. **23**, 3912–3922
54. Varghese, S., Gupta, D., Baran, T., Jiemjit, A., Gore, S.D., Casero, Jr, R.A. and Woster, P.M. (2005) Alkyl-substituted polyaminohydroxamic acids: a novel class of targeted histone deacetylase inhibitors. J. Med. Chem. **48**, 6350–6365
55. Varghese, S., Senanayake, T., Murray-Stewart, T., Doering, K., Fraser, A., Casero, Jr, R.A. and Woster, P.M. (2008) Polyaminohydroxamic acids and polyaminobenzamides as isoform selective histone deacetylase inhibitors. J. Med. Chem. **51**, 2447–2456
56. Tabe, Y., Jin, L., Contractor, R., Gold, D., Ruvolo, P., Radke, S., Xu, Y., Tsutusmi-Ishii, Y., Miyake, K., Miyake, N. et al. (2007) Novel role of HDAC inhibitors in AML1/ETO AML cells: activation of apoptosis and phagocytosis through induction of annexin A1. Cell Death Differ. **14**, 1443–1456
57. Chuthapisith, S., Bean, B.E., Cowley, G., Eremin, J.M., Samphao, S., Layfield, R., Kerr, I.D., Wiseman, J., El-Sheemy, M., Sreenivasan, T. and Eremin, O. (2009) Annexins in human breast cancer: possible predictors of pathological response to neoadjuvant chemotherapy. Eur. J. Cancer **45**, 1274–1281
58. Shi, Y., Lan, F., Matson, C., Mulligan, P., Whetstine, J.R., Cole, P.A., Casero, R.A. and Shi, Y. (2004) Histone demethylation mediated by the nuclear amine oxidase homolog LSD1. Cell **119**, 941–953
59. Wang, Y., Murray-Stewart, T., Devereux, W., Hacker, A., Frydman, B., Woster, P.M. and Casero, Jr, R.A. (2003) Properties of purified recombinant human polyamine oxidase, PAOh1/SMO. Biochem. Biophys. Res. Commun. **304**, 605–611
60. Ferioli, M.E., Berselli, D. and Caimi, S. (2004) Effect of mitoguazone on polyamine oxidase activity in rat liver. Toxicol. Appl. Pharmacol. **201**, 105–111
61. Bellelli, A., Cavallo, S., Nicolini, L., Cervelli, M., Bianchi, M., Mariottini, P., Zelli, M. and Federico, R. (2004) Mouse spermine oxidase: a model of the catalytic cycle and its inhibition by *N*,*N*1-bis(2, 3-butadienyl)-1,4-butanediamine. Biochem. Biophys. Res. Commun. **322**, 1–8
62. Wang, Y., Hacker, A., Murray-Stewart, T., Frydman, B., Valasinas, A., Fraser, A.V., Woster, P.M. and Casero, Jr, R.A. (2005) Properties of recombinant human *N*1-acetylpolyamine oxidase (hPAO): potential role in determining drug sensitivity. Cancer Chemother. Pharmacol. **56**, 83–90
63. Cona, A., Manetti, F., Leone, R., Corelli, F., Tavladoraki, P., Polticelli, F. and Botta, M. (2004) Molecular basis for the binding of competitive inhibitors of maize polyamine oxidase. Biochemistry **43**, 3426–3435

64. Huang, Y., Greene, E., Murray Stewart, T., Goodwin, A.C., Baylin, S.B., Woster, P.M. and Casero, Jr, R.A. (2007) Inhibition of lysine-specific demethylase 1 by polyamine analogues results in reexpression of aberrantly silenced genes. Proc. Natl. Acad. Sci. U.S.A. **104**, 8023–8028
65. Creaven, P.J., Perez, R., Pendyala, L., Meropol, N.J., Loewen, G., Levine, E., Berghorn, E. and Raghavan, D. (1997) Unusual central nervous system toxicity in a phase I study of *N*1,*N*11 diethylnorspermine in patients with advanced malignancy. Invest. New Drugs **15**, 227–234
66. Streiff, R.R. and Bender, J.F. (2001) Phase 1 study of *N*1-*N*11-diethylnorspermine (DENSPM) administered TID for 6 days in patients with advanced malignancies. Invest. New Drugs **19**, 29–39
67. Hahm, H.A., Ettinger, D.S., Bowling, K., Hoker, B., Chen, T.L., Zabelina, Y. and Casero, Jr, R.A. (2002) Phase I study of *N*(1),*N*(11)-diethylnorspermine in patients with non-small cell lung cancer. Clin. Cancer Res. **8**, 684–690

Essays Biochem. (2009) **46**, 95–110; doi:10.1042/BSE0460007

7

Polyamine analogues targeting epigenetic gene regulation

Yi Huang*, Laurence J. Marton†, Patrick M. Woster¶ and Robert A. Casero, Jr*[1]

**The Sidney Kimmel Comprehensive Cancer Center at Johns Hopkins, The Johns Hopkins University School of Medicine, Bunting ◊ Blaustein Cancer Research Building, 1650 Orleans Street, Baltimore, MD 21231, U.S.A., †Progen Pharmaceuticals, Redwood City, CA 94065, U.S.A., and ¶Department of Pharmaceutical Sciences, Wayne State University, Detroit, MI 48202, U.S.A.*

Abstract

Over the past three decades the metabolism and functions of the polyamines have been actively pursued as targets for antineoplastic therapy. Interactions between cationic polyamines and negatively charged nucleic acids play a pivotal role in DNA stabilization and RNA processing that may affect gene expression, translation and protein activity. Our growing understanding of the unique roles that the polyamines play in chromatin regulation, and the discovery of novel proteins homologous with specific regulatory enzymes in polyamine metabolism, have led to our interest in exploring chromatin remodelling enzymes as potential therapeutic targets for specific polyamine analogues. One of our initial efforts focused on utilizing the strong affinity that the polyamines have for chromatin to create a backbone structure, which could be combined with active-site-directed inhibitor moieties of HDACs (histone deacetylases). Specific PAHAs (polyaminohydroxamic acids) and PABAs (polyaminobenzamides) polyamine analogues have demonstrated potent inhibition of the HDACs, re-expression of p21 and significant inhibition

[1]*To whom correspondence should be addressed (email rcasero@jhmi.edu).*

of tumour growth. A second means of targeting the chromatin-remodelling enzymes with polyamine analogues was facilitated by the recent identification of flavin-dependent LSD1 (lysine-specific demethylase 1). The existence of this enzyme demonstrated that histone lysine methylation is a dynamic process similar to other histone post-translational modifications. LSD1 specifically catalyses demethylation of mono- and di-methyl Lys^4 of histone 3, key positive chromatin marks associated with transcriptional activation. Structural and catalytic similarities between LSD1 and polyamine oxidases facilitated the identification of biguanide, bisguanidine and oligoamine polyamine analogues that are potent inhibitors of LSD1. Cellular inhibition of LSD1 by these unique compounds led to the re-activation of multiple epigenetically silenced genes important in tumorigenesis. The use of these novel polyamine-based HDAC or LSD1 inhibitors represents a highly promising and novel approach to cancer prevention and therapy.

Introduction

Polyamines are naturally occurring polycationic alkylamines that are essential for eukaryotic cell growth. By virtue of their positively charged amine groups, polyamines interact with negatively charged DNA, RNA, proteins and phospholipids to change their structure and conformation. The enzymes controlling polyamine metabolism and intracellular concentrations are highly regulated and can rapidly react to changing environmental conditions. Intracellular polyamine levels and metabolism are frequently dysregulated in cancer and other hyperproliferative diseases, thus making polyamine function and metabolism attractive targets for therapeutic intervention [1,2]. The key polyamine biosynthetic enzyme, ODC (ornithine decarboxylase), has long been thought to be a marker of carcinogenesis and tumour progression [3]. Inhibiting polyamine biosynthesis by specifically targeting ODC as an anticancer strategy has yet to demonstrate significant clinical success, but it has demonstrated considerable promise as a strategy for cancer chemoprevention [4]. Recently, more focus has been directed towards the development of polyamine analogues designed to mimic the regulatory roles of natural polyamines but to have altered function. A number of synthetic polyamine analogues have exhibited encouraging effects against tumour growth in both cell culture and animal studies and several hold promise as chemotherapeutic agents [5].

There are considerable data demonstrating that chromatin is a major target for the natural polyamines and polyamine-based drugs [6–8]. Therefore we have attempted to use this property to advance the hypothesis that specific polyamine analogues could target the chromatin remodelling enzymes, including the HDACs (histone deacetylases) and the newly identified histone LSD1 (lysine-specific demethylase 1). These enzymes, among others, are responsible for normal gene regulation, and in a variety of disease processes their activity may lead to aberrant silencing of important tumour suppressor genes. As aberrant epigenetic silencing of tumour suppressor genes is a common occurrence in the development of cancer, this strategy holds considerable promise for the treatment of neoplastic disease, and the present chapter will discuss the most recent findings in the field [9].

Polyamine metabolism

Polyamines are critical for eukaryotic cell growth and thus maintenance of appropriate intracellular concentrations via a highly regulated interplay between biosynthesis, catabolism, uptake and excretion is required for normal function (Figure 1). Two major regulatory enzymes of polyamine biosynthesis are ODC and AdoMetDC (*S*-adenosylmethionine decarboxylase). The regulatory protein Az (antizyme) facilitates the degradation of ODC and down-regulates the transport of polyamines into the cell, and thus is considered to be dedicated principally to the feedback regulation of

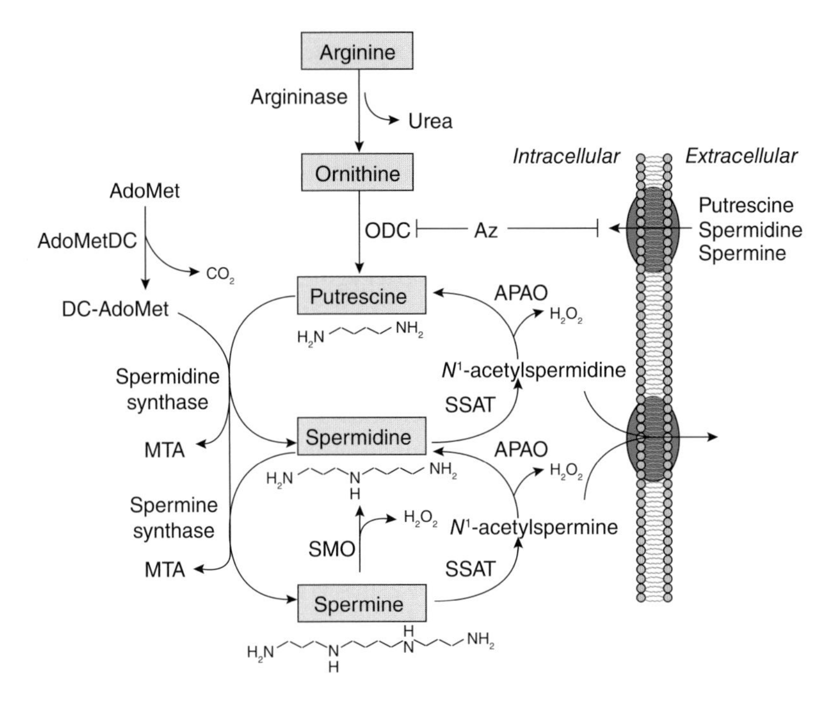

Figure 1. Polyamine metabolism in mammals

The major regulatory enzymes of polyamine biosynthesis are ODC and AdoMetDC. ODC forms putrescine from L-ornithine and is the first rate-limiting step in polyamine biosynthesis. The regulatory protein Az can facilitate degradation of ODC and negatively regulate the eukaryotic polyamine transport system. AdoMetDC forms decarboxylated *S*-adenosylmethionine (DC-AdoMet) from *S*-adenosylmethionine, and DC-AdoMet acts as an aminopropyl donor. Spermidine synthase and spermine synthase transfer the aminopropyl group from DC-AdoMet to putrescine or spermidine for the synthesis of spermidine, spermine and MTA (5′-deoxy-5′-m ethylthioadenosine) respectively. After acetylation of spermine and spermidine by SSAT to form N^1-acetylspermine and N^1-acetylspermidine respectively, acetyl derivatives are then cleaved into 3-aceto-aminopropanal, the ROS H_2O_2 and spermidine and putrescine respectively, through the action of FAD-dependent APAO. SMO is a highly inducible FAD-dependent enzyme that directly oxidizes spermine to produce spermidine, 3-aminopropanal and H_2O_2.

polyamine levels [10]. Polyamine catabolism was initially thought to be a two-step procedure regulated by the rate-limiting enzyme SSAT (spermidine/spermine N^1-acetyltransferase), which provides substrate for the generally constitutively expressed APAO (N^1-acetylpolyamine oxidase) [11]. After acetylation of spermine and spermidine by SSAT to form N^1-acetylspermine and N^1-acetylspermidine respectively, acetyl derivatives are then cleaved into 3-acetamidopropanal, the ROS (reactive oxygen species) H_2O_2, and spermidine and putrescine respectively, through the action of FAD-dependent APAO. However, recent studies from our laboratory and others have demonstrated that a variably spliced human SMO (spermine oxidase) efficiently uses unacetylated spermine as a substrate, and that this enzyme is inducible by specific polyamine analogues [12–15]. These findings indicate the existence of a more complex polyamine catabolic pathway in the regulation of polyamine homoeostasis than originally thought.

Epigenetic regulation of gene expression

Epigenetic modification of chromatin is a major regulator of gene expression. Epigenetic modifications refer to heritable changes in chromatin/DNA that are not due to changes in primary sequence. These include a variety of modifications of the histone proteins and methylation of cytosine residues in DNA. The histone proteins are the major packaging proteins of eukaryotic DNA. The eukaryotic genome requires packaging of DNA both for structural purposes, as a means of fitting 2 m of DNA into the nucleus, and as one mechanism for the regulation of gene expression. Two of each of the core histone proteins, H2A, H2B, H3 and H4, are assembled into the nucleosome, around which 146 bp of DNA are wound. This nucleosome is the basic packaging unit of eukaryotic chromatin and the density of nucleosomes and the affinity of the individual nucleosomes on any stretch of DNA determine the accessibility of the gene to various factors, including the basic transcriptional machinery. Importantly, in cancer, aberrant epigenetic silencing of gene expression, including tumour suppressor genes, is a common occurrence [9].

There are numerous possible modifications of histone tails, including acetylation, methylation, ubiquitination, phosphorylation, SUMOylation and ribosylation, each of which can affect the expression of genes [16]. The specific modification of histones determines, in part, which regions of the genome are in an open and transcriptionally active confirmation and which are closed and transcriptionally inactive. The most studied of the histone modifications are acetylation/deacetylation and methylation. Acetylated histones are typically associated with active gene transcription. However, histone methylation can either be activating or inhibitory with respect to transcription, depending on the specific residue methylated. Histones may be methylated on either lysine (K) or arginine (R) residues, and the methylation status of key histone lysine residues, such as H3K4, H3K9, H3K27, H3K36, H3K79 and H4K20, represent specific epigenetic marks that are associated with transcriptional regulation. Although the dynamic nature of histone acetylation, mediated through the counterbalancing activity of HATs (histone acetyltransferases) and HDACs, has been known for some time [17], only recently has a similar dynamic regulation of histone methylation been demonstrated. Shi et al. [18] reported the discovery of the first enzyme able to specifically demethylate Lys[4] of histone H3. The newly identified protein was therefore named LSD1 (lysine-specific demethylase; BHC110) [18]. Subsequent to the discovery of LSD1, a family of Jmj C (Jumonji C) domain-containing histone demethylases was also identified, thus expanding the number of known proteins involved in chromatin remodelling [19]. These findings firmly establish that histone methylation is a dynamic process under enzymatic control, similar to the other post-translational histone modifications, and suggest that this modifying enzyme may also be a rational target for therapeutic intervention [20].

Exploiting polyamine structure to target aberrantly silenced genes

The incorporation of a polyamine structure into the design of a putative drug offers several benefits as a targeting vehicle, particularly when chromatin is the target. A selective, energy-dependent polyamine transport system allows molecules resembling the general polyamine structure to be accumulated by cells, and the cationic nature of the polyamine backbone provides affinity to the negatively charged chromatin. Several investigators have used the polyamine structure as a backbone to which various active moieties have been conjugated, including alkylating agents, DNA intercalators and other antiproliferative agents [21,22]. Delcros et al. [23] have tested the limits of polyamine transport with respect to size and types of molecules that can be attached to the polyamine backbone and still be effectively transported. The findings of each of the above studies clearly indicate that the polyamine structure can be used effectively to transport specific active moieties into cells and in many cases, target chromatin. We recently attempted to build on this paradigm with a new generation of agents that target chromatin [25,26,36]. However, unlike previous attempts, our goal was not to damage chromatin, but rather to alter its regulation of gene expression.

Polyamine analogues as HDAC inhibitors

The growing interest in drugs that alter chromatin is based on the recognition that epigenetic regulation of gene expression plays a critical role in the aetiology and progression of cancer and, unlike gene mutations or loss, epigenetic changes are, in theory, reversible [9]. Aberrant silencing of tumour suppressors and other genes has been found in all cancers examined. One of the major regulators of chromatin conformation, and hence gene expression, is histone acetylation. As stated above, acetylation of histones is generally associated with transcriptionally active chromatin, and activity of HDACs lead to condensation of chromatin and inhibition of transcription. A number of class I/II HDAC inhibitors have been examined for their ability to re-express silenced genes [24]. Although several class I/II HDCA inhibitors have been synthesized and some are currently in clinical trial, virtually all of them have been designed to target primarily the zinc cofactor at the active site of the HDACs [24].

We have previously reported the use of various polyamine analogue structures combined with active-site-directed inhibitor moieties of the class I/II HDACs [25,26]. This strategy has the advantage of using known structures that inhibit the HDACs, the hydroxamic acids and benzamides, combined with the high chromatin affinity of the polyamine structure. Additionally, as tumour cells are known to transport and accumulate many of the polyamine analogues, the polyamine backbone offers the likelihood that these compounds would be readily transported into tumour cells. Specific members of the PAHA (polyaminohydroxamic acid) and PABA (polyaminobenzamide) polyamine analogues demonstrated potent inhibition of the class I/II HDACs in a cell-free system, and in treated leukaemia cells re-expression of the growth regulating CDK (cyclin-dependent kinase) inhibitor p21 was observed. Some of these analogues also significantly inhibited tumour cell growth *in vitro*.

An additional advantage of using the polyamines as a backbone for HDAC inhibitors is that it provides a scaffold upon which it is possible to design molecules that possess selectivity for the individual HDACs (see Chapter 6 by Woster). This possibility would be useful in the design of molecular tools to study the effects of inhibiting the individual HDACs, and may provide a therapeutic advantage over existing non-selective inhibitors. One promising compound that demonstrates some selectivity with respect to HDAC inhibition was the compound designated **17** (Figure 2) in Varghese et al. [26]. The class II HDAC6, in addition to having histone targets, is also capable of deacetylating α-acetyltubulin. Compound **17** demonstrated potent functional *in situ* inhibition of HDAC6 resulting in a substantial increase in α-acetyltubulin in treated cells. These data underscore the possibility of using the flexibility allowed by the polyamine structure to design selective inhibitors for each of the individual class I/II HDACs.

Although considerable work remains to be done, the initial analysis of the polyamine analogue HDAC inhibitors of both the PAHA and PABA families shows considerable promise.

Targeting LSD1 for gene re-expression

As stated above, the discovery of LSD1 and the Jmj C domain-containing demethylases indicated that histone methylation, like histone acetylation, is a dynamic process. Structural analysis demonstrates that LSD1 is highly conserved across species and consists of an N-terminal SWIRM (Swi3p/Rsc8c/Moira) domain, a central protruding tower domain and a C-terminal amine oxidase-like domain, containing a FAD-binding subdomain, which is highly homologous with MAOs (monoamine oxidases) and the polyamine oxidases,

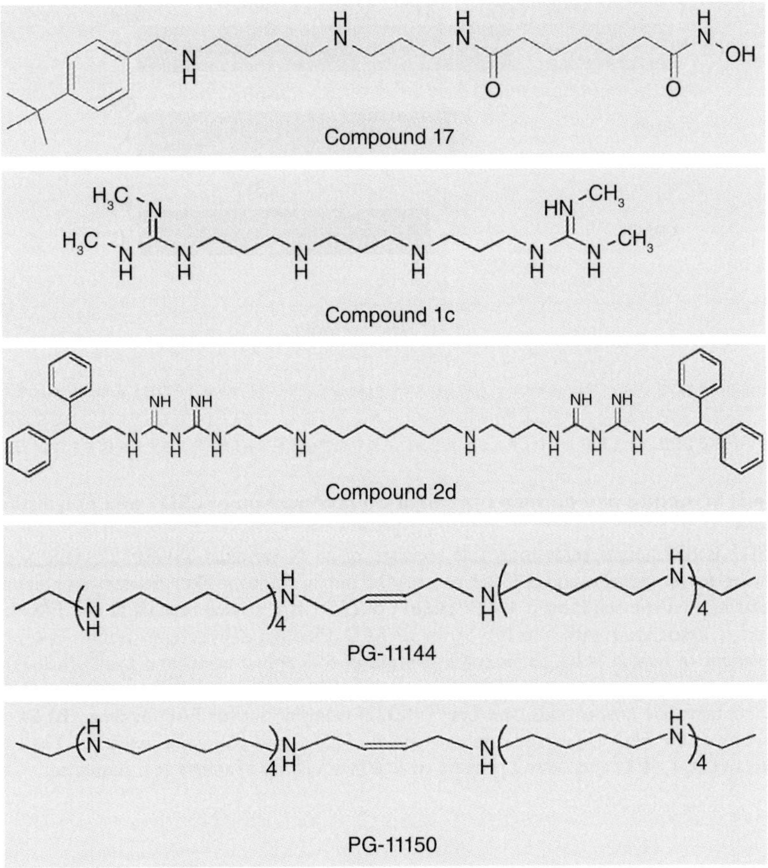

Compound 17

Compound 1c

Compound 2d

PG-11144

PG-11150

Figure 2. Chemical structures of polyamine analogues
Compound **17** selectively inhibits HDAC6 activity and increases acetylated α-tubulin in HCT116 colorectal cancer cells. **1c** and **2d** are potent inhibitors of LSD1 activity and re-activate aberrantly silenced genes in tumour cells. PG-11144 and PG-11150 polyamine analogues contain ten amines and are a *cis/trans* pair with double bonds in the centre of their structure. Oligoamines competitively inhibit LSD1 activity and re-activate aberrantly silenced genes in colorectal cancer cells.

SMO and APAO [18,27] (Figure 3). LSD1 catalyses the demethylation of mono- or di-methylated Lys[4] of histone H3 (H3K4) by cleavage of the α-carbon bond of the substrate through an oxidative process with the reduction of FAD. $FADH_2$ is re-oxidized by oxygen to produce H_2O_2, leading to the generation of an imine intermediate. The imine intermediate is subsequently hydrolysed to generate formaldehyde and the amine of lysine (Figure 4). Despite the similarity of structure and chemistry between these amine oxidases, LSD1 demonstrates entirely different biological activity and cellular localization than do the MAOs (Table 1). MAOs, including two isoforms MAO A and MAO B, are bound to the outer mitochondrial membrane [28]. APAO is a peroxisomal enzyme, whereas SMO is found in both the cytoplasm and nucleus [14]. Both APAO and SMO are FAD-dependent enzymes

A

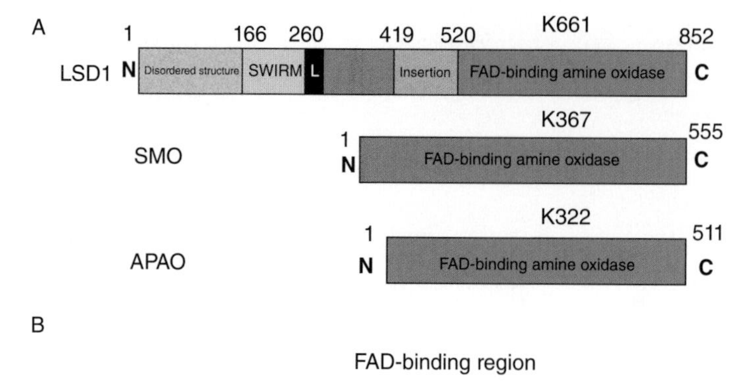

B

FAD-binding region

LSD1 635 P A V Q F V P P L P E W K T S A V Q R M G F G N L N K V V L C F D R V F W D P S V 675
SMO 342 T S F – F R P G L P T E K V A A I H R L G I G T T D K I F L E E E E P F W G P E C 381
APAO 296 L D T F F D P P L P A E K A E A I R K I G F G T N N K I F L E F E E P F W E P D C 336

Figure 3. Structure and domain organization comparison of LSD1 and polyamine oxidases
(**A**) LSD1 is 852 amino acids long and consists of an N-terminal SWIRM domain, a central protruding tower domain and a C-terminal AOL (amine oxidase-like) domain that contains a FAD-binding catalytic subdomain. Lys^{661} (K661) of LSD1 is a critical residue at the FAD-binding site that is associated with the N5 atom of FAD through a hydrogen bond. The primary splice variant of human SMO codes for a protein of 511 amino acids, and Lys^{367} (K367) at the C-terminal is the key residue for FAD association. The predominant human splice variant of APAO contains 551 amino acids, and Lys^{322} (K322) is important for FAD binding. (**B**) Structural comparison of the FAD-associated catalytic centre of LSD1 and polyamine oxidases. The catalytic domains of LSD1, SMO and APAO possess over 60% similarity in amino acid sequences.

that oxidize polyamine substrates and produce H_2O_2. The structural and catalytic similarities of these FAD-dependent oxidases has been instructive in understanding the basic biology of FAD-dependent oxidases, as well as in the search for effective inhibitors that can interact with LSD1.

H3K4
R–NH
$^+$ CH$_3$
LSD1
FAD
FADH$_2$
H$_2$O$_2$
O$_2$
H3K4
R–N
$^+$ = CH$_2$
H$_2$O
R = H or CH$_3$
H3K4
R–N
CH$_2$OH
O
C
H H
Formaldehyde
H3K4
R–NH$_2$
$^+$
Di- or mono-methyl H3K4
Imine intermediate
Carbinolamine
Mono- or non-methyl H3K4

Figure 4. Mechanism of histone demethylation by LSD1
LSD1 catalyses the FAD-dependent demethylation of mono- (R=H) or di-methyl ($R=CH_3$) Lys^4 of histone 3 through transferring two hydrogen atoms from methylated H3K4 to FAD with the resultant reduction of oxygen to H_2O_2. The imine intermediate is then hydrolysed to an unstable carbinolamine that subsequently degrades to release formaldehyde.

Table 1. Comparison of biological functions between amine oxidases and LSD1

Characteristic	MAO	APAO	SMO	LSD1
Subcellular localization	Mitochondrial membrane	Peroxisome	Cytoplasm, nucleus	Nucleus
Substrate	Arylalkyl amines	N^1-acetyl spermine, N^1-acetyl spermidine	Spermine, spermidine	Histone, proteins
FAD-dependent	Yes	Yes	Yes	Yes
H_2O_2 product	Yes	Yes	Yes	Yes
Aldehyde product	Yes	Yes	Yes	Yes
Cellular function	Neurological activities	Polyamine catabolism	Polyamine homoeostasis	Transcription regulation

Linking LSD1-suppressed gene expression to the role of LSD1 in tumorigenesis

LSD1 is a member of a multiprotein co-repressor complex that includes CoRest, HDAC1/2 and BHC80. Several recent studies demonstrate that inhibition of LSD1 by siRNA (small interfering RNA) or pharmacological inhibition changes global and promoter-specific H3K4me2 (dimethylated H3K4) levels and increases expression of some known LSD1/CoREST-target genes. For example, Shi et al. [18] demonstrated an up-regulation of neuron-specific M4 AchR, SCN1A-3A and the CDK inhibitor $p57^{KIP2}$ in LSD1 RNAi (RNA interference)-treated HeLa cells. Lee et al. [27] reported an increase in global H3K4me2, as well as transcriptional de-repression of two LSD1 target genes, Egr1 and the pluripotent stem cell marker Oct4, in P19 EC cells treated with a non-specific MAO inhibitor. A recent study demonstrated that knockout of LSD1 in embryonic stem cells induces progressive loss of DNA methylation and a decrease in DNA methyltransferase 1 stability and protein levels [29]. These interesting findings demonstrate a previously unknown mechanistic interaction between histone demethylase and DNA methyltransferase.

DNA promoter hypermethylation frequently acts in concert with histone modifications that result in decreased chromatin-activating marks, such as H3K4me (monomethylated H3K4), resulting in increased repressive marks such as H3K9me and H3K27me, and in the aberrant silencing of specific genes [30]. Since the discovery of LSD1 as an important demethylase of the key activating mark H3K4, the potential association of LSD1 activity with tumorigenesis has been intensively investigated. A number of recent studies indicate that an alteration in the function of LSD1 has a role in cancer. For example, high-level LSD1 expression has been linked to an increased risk of prostate cancer recurrence, suggesting a tumour-promoting role for LSD1 [31]. In another study, LSD1 was shown to possess a pro-oncogenic activity through its regulation of pro-survival gene expression and p53 transcription in human breast cancer cells [32]. Bradley et al. [33] reported that induction of LSD1 is one of the early responses to chemical carcinogens in HMECs (human mammary epithelial cells), and that it may affect the expression of multiple genes critical in early-stage mammary oncogenic transformation. Very recently, Schulte et al. [34], demonstrated that a non-specific MAO inhibitor is effective in inducing expression of genes associated with differentiation in neuroblastoma cells, and that the inhibitor can cause significant *in vivo* inhibition of tumour growth. These studies have established LSD1 as an important link to the development and progression of cancer and provide a rationale for developing LSD1 inhibitors as a target for therapeutic intervention.

Identification of polyamine analogues as effective LSD1 inhibitors.
Human LSD1 shares 20% similarity in overall structure with that of other FAD-dependent amine oxidases. Specifically, the catalytic domains of LSD1 and SMO are over 60% similar in amino acid sequence (Figure 3). This similarity suggests that specific polyamine analogues may function as effective inhibitors of LSD1 (Table 2).

Although the natural polyamines or acetylpolyamines are not substrates of LSD1 [18], the strong association of polyamines with chromatin and the structural similarity between the polyamines and the lysine tails of histones suggest that polyamines and/or polyamine analogues may alter the activity of LSD1 and other chromatin modifiers. In our previous studies we discovered that specific polyamine analogues function as potent inhibitors of purified polyamine oxidases [15,35]. Since the active site of LSD1 is closely related to that of SMO and APAO, and because guanidines have been shown to inhibit the activity of structurally related polyamine oxidases, we sought to determine whether a small library of bisguanidine and biguanide polyamine analogues (Figure 2) functioned as effective inhibitors of LSD1. Most of these compounds were found to inhibit the demethylase activity of recombinant LSD1 by >50% at concentrations less than 1 μM [36]. The two most

Table 2. Inhibitory effects of polyamine analogues on the activity of recombinant polyamine oxidases, LSD1 and tumour cell growth

–, No effect; + to +++, modest to potent inhibition.

Polyamine	APAO	SMO	LSD1	Growth
1c	++	+/–	+++	+
2d	+++	+++	+++	++
PG-11144	Unknown	+++	++	+++
PG-11150	Unknown	+++	++	+++

potent analogues, **1c** [1,11-bis(N^2,N^3-dimethyl-N^1-guanidino)-4,8-diaza-undecane] and **2d** (1,15-bis{N^5-[3,3-(diphenyl)propyl]-N^1-biguanido}-4,12-di-azapentadecane), exhibited non-competitive inhibition kinetics, suggesting that these compounds probably bind to LSD1 at a site other than at the active site. The possible mechanisms of action include the fact that these LSD1 inhibitors may block FAD from associating with its binding site, or the fact that these inhibitors might induce a conformational change in LSD1, such that LSD1 is no longer able to bind to its histone lysine substrate correctly. It is, however, possible that the enzymatic kinetics obtained when using recombinant LSD1 and a short peptide as a substrate in a cell-free system may not be representative of the interaction of the inhibitor and the nucleosome in association with the LSD1–CoRest–HDAC complex *in situ*.

A class of long-chain polyamine analogues known as oligoamines has also been found to potently inhibit APAO or SMO (Table 2), suggesting that these analogues may also inhibit LSD1 [36]. Among others, we tested two *cis/trans* oligoamine isomers, PG-11144 and PG-11150 (Figure 2), for their ability to inhibit LSD1 [36] These oligoamines exhibit competitive-inhibition kinetics, suggesting that the oligoamines may directly compete with substrate at the active site in a manner different than seen for the bisguanidine and biguanide analogues. This new class of polyamine analogue LSD1 inhibitors offers another promising avenue of investigation.

It should be noted that the precise mechanism of inhibition of LSD1 by the polyamine analogues remains unclear. Ongoing crystallographic analysis should facilitate a better understanding of the molecular and structural basis of LSD1 inhibition by polyamine analogues, and perhaps provide insight into the design of more effective inhibitors [37].

Inhibition of LSD1 by novel polyamine analogues reactivates silenced genes in cancer cells

To determine whether the *in vitro* inhibition of LSD1 activity by polyamine analogues translated into a cellular response, the effects of polyamine analogues as LSD1 inhibitors on global H3K4me status were examined in multiple human cancer cell lines, including colorectal, breast, lung and leukaemia. Exposure of cancer cells to these polyamine analogues leads to increased global H3K4me2 and H3K4me1, with no change in H3K4me3 (trimethylated H3K4) and H3K9me2 [36,38]. The promoter region of H3K4me2 is usually associated with open chromatin and active transcription, and the occupancy of H3K4me2 was found to be at low levels in the promoters of a number of frequently DNA hypermethylated and epigenetically silenced genes important in tumorigenesis [30]. A number of such silenced genes have been identified in colorectal cancer cell lines, such as HCT116 and RKO. These genes include members of the WNT signalling pathway antagonists, SFRPs (secreted frizzle-related protein) family, GATA family transcription factors, the mismatch repair gene *MLH1*, the cell-cycle regulator *CDKN2A*, and the tissue invasion regulator *TIMP3* (tissue inhibitor of metalloproteinase 3) [30]. Treatment of HCT116 cells with **1c** or **2d** for 48 h resulted in the re-expression of three members of the SFRP family, *SFRP1*, *SFRP4* and *SFRP5*, as well as of the *GATA5* transcription factor [36]. Similarly, treatment with PG-11144 or PG-11150 for 24 h led to the re-expression of *SFRP1* and *SFRP2* [38]. Leukaemia cells treated with **2d** resulted in increased global H3K4me2 and the re-expression of E-cadherin, an important tumour suppressor gene [39].

Chromatin immunoprecipitation analysis demonstrated that polyamine analogue-induced gene re-expression in HCT116 cells is closely associated with increased active histone marks H3K4me1/me2 and acetyl-H3K9, and decreased repressive marks H3K9me1/me2 [36,38]. However, no change of H3K4me3 levels were detected after treatment with the polyamine analogues, suggesting that these compounds specifically target LSD1 rather than other histone demethylases such as the Jmj C histone demethylases that are capable of demethylating H3K4me3 [19]. Some other hallmarks of silenced chromatin, such as H3K9me3 and PcG (polycomb) group protein-mediated H3K27 methylation, remain unchanged in the promoter regions of the analogue-activated genes after treatment with LSD1 inhibitors. These results are similar to the chromatin state of HCT116 treated with the DNA methylation inhibitor DAC (5-aza-2′-deoxycytidine) [40].

As noted above, CpG island hypermethylation frequently acts in concert with abnormal histone mark activities in silencing genes. Interestingly, the re-expression of aberrently silenced genes by our unique polyamine analogues occurs without wholesale changes in the methylation status of CpG islands of the gene examined. Similar results were observed in silenced genes that were reactivated by inhibition of the class III HDAC, SIRT1 (sirtuin 1) [41]. These results indicate that treatment with LSD1 inhibitors alone is sufficient to produce gene re-expression, even when dense promoter DNA methylation is maintained.

It is possible that the polyamine analogues have effects other than inhibition of LSD1 that lead to gene re-expression. To address this issue, we compared the effects of the inhibition of LSD1 with the analogues **1c** and **2d** with those induced by an RNAi-mediated knockdown in LSD1 expression in HCT116 cells. LSD1 depletion by RNAi was accompanied by increased H3K4me2 at the promoters of re-activated genes, and resulted in modest re-expression of the examined genes [36]. We found that pharmacological inhibition of LSD1 with the analogues was more effective than was RNAi with respect to re-expression of silenced genes, which may reflect inherent differences in chromatin structure resulting from inhibitor–LSD1 complexes compared with RNAi-induced decreases in LSD1 protein.

Combination of LSD1 inhibitors with other agents targeting epigenetic regulation of gene expression

Combination chemotherapy is an important strategy in modern cancer treatment, frequently having the advantage of allowing for lower and better-tolerated doses of individual chemotherapeutic agents, while at the same time improving the overall response as compared with when the agents are used alone. The combination of DNA methyltransferase and HDAC inhibitors has demonstrated synergistic effects in re-expressing epigenetically silenced genes in cultured cancer cells, and has produced clinical responses in patients with leukaemias [42,43]. Our latest studies demonstrate that the combination of low-dose oligoamines with DAC results in synergistic expression of the *SFRP2* gene [38]. This result suggests that the combination of LSD1 inhibitors and DNA methylation inhibitors can collaborate in the re-expression of specific silenced genes and may provide a useful strategy for the potential clinical utility of these two agents, each of which targets different epigenetic regulatory enzymes.

Conclusions

Increasing knowledge regarding the molecular mechanisms of polyamine metabolism in cancer cells that has accumulated over the last three decades has provided unique opportunities for the development of many promising therapeutic agents. Despite the somewhat disappointing results obtained from clinical trials of the early polyamine biosynthesis inhibitors, polyamine metabolism remains a rational target for cancer therapy, and opportunities still exist for development of novel agents with better specificity when targeting cancer cells and fewer side effects affecting normal tissues. The unique association of 'superinduction' of polyamine catabolic enzymes with a cytotoxic response for certain polyamine analogues has been well elucidated and noted. The interaction of polyamine or polyamine-based analogues with nucleic acids has received considerable attention, although the exact mechanisms of this phenomenon are still being defined. Epigenetic regulation of gene expression has emerged as an important target for drug development in cancer therapy. Specific polyamine analogues have been identified as potent inhibitors of the HDACs. LSD1 is a newly identified homologue of polyamine oxidase. Owing to its important activity in modifying chromatin and gene transcription, significant attention has been focused on developing LSD1 inhibitors as an effective strategy in treatment of cancer and other diseases. Our recent findings clearly suggest that novel polyamine analogues are powerful inhibitors of LSD1, both alone and in combination with other agents that target epigenetic silencing, and that they are capable of inducing re-expression of several aberrantly silenced genes important in tumorigenesis. The use of these compounds represents a new direction for drug development in cancer prevention and therapy.

Summary

- *Polyamine effects on chromatin remodelling are important in the stabilization of DNA structure and in RNA processing and may affect diverse cellular functions including transcriptional regulation.*

- *Drug design based on polyamine structure, when combined with active site-directed inhibitor moieties of the HDACs, led to the successful development of specific HDAC inhibitors of the PAHA and PABA families. These inhibitors demonstrate potent inhibition of HDAC in tumour cells.*
- *The recent identification of flavin-dependent LSD1, which specifically demethylates mono- and di-methyl Lys⁴ of histone 3, plays critical roles in regulating gene transcription.*
- *The structural and catalytic similarities of LSD1 and polyamine oxidases have facilitated the identification of a group of biguanide, bisguanidine and oligoamine polyamine analogues as potent inhibitors of LSD1.*
- *Inhibition of LSD1 alone and in combination with other agents targeting epigenetic silencing are capable of inducing re-expression of a number of aberrantly silenced genes that are important in tumorigenesis.*

References

1. Casero, Jr, R.A. and Marton, L.J. (2007) Targeting polyamine metabolism and function in cancer and other hyperproliferative diseases. Nat. Rev. Drug Discov. **6**, 373–390
2. Wallace, H.M., Fraser, A.V. and Hughes, A. (2003) A perspective of polyamine metabolism. Biochem. J. **376**, 1–14
3. Pegg, A.E. (2006) Regulation of ornithine decarboxylase. J. Biol. Chem. **281**, 14529–14532
4. Gerner, E.W. and Meyskens, Jr, F.L. (2009) Combination chemoprevention for colon cancer targeting polyamine synthesis and inflammation. Clin. Cancer Res. **15**, 758–761
5. Seiler, N. (2005) Pharmacological aspects of cytotoxic polyamine analogs and derivatives for cancer therapy. Pharmacol. Ther. **107**, 99–119
6. Basu, H.S., Smirnov, I.V., Peng, H.F., Tiffany, K. and Jackson, V. (1997) Effects of spermine and its cytotoxic analogs on nucleosome formation on topologically stressed DNA *in vitro*. Eur. J. Biochem. **243**, 247–258
7. Basu, H.S., Sturkenboom, M.C., Delcros, J.G., Csokan, P.P., Szollosi, J., Feuerstein, B.G. and Marton, L.J. (1992) Effect of polyamine depletion on chromatin structure in U-87 MG human brain tumour cells. Biochem. J. **282**, 723–727
8. Sato, N., Ohtake, Y., Kato, H., Abe, S., Kohno, H. and Ohkubo, Y. (2003) Effects of polyamines on histone polymerization. J. Protein Chem. **22**, 303–307
9. Jones, P.A. and Baylin, S.B. (2007) The epigenomics of cancer. Cell **128**, 683–692
10. Coffino, P. (2001) Regulation of cellular polyamines by antizyme. Nat. Rev. Mol. Cell Biol. **2**, 188–194
11. Pegg, A.E. (2008) Spermidine/spermine-N^1-acetyltransferase: a key metabolic regulator. Am. J. Physiol. Endocrinol. Metab. **294**, E995–E1010
12. Wang, Y., Devereux, W., Woster, P., Stewart, T., Hacker, A. and Casero, Jr, R. (2001) Cloning and characterization of a human polyamine oxidase that is inducible by polyamine analogue exposure. Cancer Res. **61**, 5370–5373
13. Vujcic, S., Diegelman, P., Bacchi, C.J., Kramer, D.L. and Porter, C.W. (2002) Identification and characterization of a novel flavin-containing spermine oxidase of mammalian cell origin. Biochem. J. **367**, 665–675
14. Murray-Stewart, T., Wang, Y., Goodwin, A., Hacker, A., Meeker, A. and Casero, Jr, R.A. (2008) Nuclear localization of human spermine oxidase isoforms: possible implications in drug response and disease etiology. FEBS J. **275**, 2795–2806

15. Wang, Y., Murray-Stewart, T., Devereux, W., Hacker, A., Frydman, B., Woster, P. and Casero, Jr, R. (2003) Properties of purified recombinant human polyamine oxidase, PAOh1/SMO. Biochem. Biophys. Res. Commun. **304**, 605–611
16. Jenuwein, T. and Allis, C.D. (2001) Translating the histone code. Science **293**, 1074–1080
17. Kuo, M.H. and Allis, C.D. (1998) Roles of histone acetyltransferases and deacetylases in gene regulation. BioEssays **20**, 615–626
18. Shi, Y., Lan, F., Matson, C., Mulligan, P., Whetstine, J., Cole, P., Casero, R. and Shi, Y. (2004) Histone demethylation mediated by the nuclear amine oxidase homolog LSD1. Cell **119**, 941–953
19. Klose, R.J., Kallin, E.M. and Zhang, Y. (2006) JmjC-domain-containing proteins and histone demethylation. Nat. Rev. Genet. **7**, 715–727
20. Shi, Y. (2007) Histone lysine demethylases: emerging roles in development, physiology and disease. Nat. Rev. Genet. **8**, 829–833
21. Holley, J.L., Mather, A., Wheelhouse, R.T., Cullis, P.M., Hartley, J.A., Bingham, J.P. and Cohen, G.M. (1992) Targeting of tumor cells and DNA by a chlorambucil-spermidine conjugate. Cancer Res. **52**, 4190–4195
22. Phanstiel, O.I., Price, H.L., Wang, L., Juusola, J., Kline, M. and Shah, S.M. (2000) The effect of polyamine homologation on the transport and cytotoxicity properties of polyamine-(DNA-intercalator) conjugates. J. Org. Chem. **65**, 5590–5599
23. Delcros, J.G., Tomasi, S., Duhieu, S., Foucault, M., Martin, B., Le Roch, M., Eifler-Lima, V., Renault, J. and Uriac, P. (2006) Effect of polyamine homologation on the transport and biological properties of heterocyclic amidines. J. Med. Chem. **49**, 232–245
24. Bolden, J.E., Peart, M.J. and Johnstone, R.W. (2006) Anticancer activities of histone deacetylase inhibitors. Nat. Rev. Drug Discov. **5**, 769–784
25. Varghese, S., Gupta, D., Baran, T., Jiemjit, A., Gore, S.D., Casero, Jr, R.A. and Woster, P.M. (2005) Alkyl-substituted polyaminohydroxamic acids: a novel class of targeted histone deacetylase inhibitors. J. Med. Chem. **48**, 6350–6365
26. Varghese, S., Senanayake, T., Murray-Stewart, T., Doering, K., Fraser, A., Casero, Jr, R.A. and Woster, P.M. (2008) Polyaminohydroxamic acids and polyaminobenzamides as isoform selective histone deacetylase inhibitors. J. Med. Chem. **51**, 2447–2456
27. Lee, M., Wynder, C., Schmidt, D., McCafferty, D. and Shiekhattar, R. (2006) Histone H3 lysine 4 demethylation is a target of nonselective antidepressive medications. Chem. Biol. **13**, 563–567
28. Binda, C., Mattevi, A. and Edmondson, D.E. (2002) Structure–function relationships in flavoenzyme-dependent amine oxidations: a comparison of polyamine oxidase and monoamine oxidase. J. Biol. Chem. **277**, 23973–23976
29. Wang, J., Hevi, S., Kurash, J.K., Lei, H., Gay, F., Bajko, J., Su, H., Sun, W., Chang, H., Xu, G. et al. (2009) The lysine demethylase LSD1 (KDM1) is required for maintenance of global DNA methylation. Nat. Genet. **41**, 125–129
30. Baylin, S.B. and Ohm, J.E. (2006) Epigenetic gene silencing in cancer: a mechanism for early oncogenic pathway addiction? Nat. Rev. Cancer **6**, 107–116
31. Kahl, P., Gullotti, L., Heukamp, L.C., Wolf, S., Friedrichs, N., Vorreuther, R., Solleder, G., Bastian, P.J., Ellinger, J., Metzger, E. et al. (2006) Androgen receptor coactivators lysine-specific histone demethylase 1 and four and a half LIM domain protein 2 predict risk of prostate cancer recurrence. Cancer Res. **66**, 11341–11347
32. Scoumanne, A. and Chen, X. (2007) The lysine-specific demethylase 1 is required for cell proliferation in both p53-dependent and -independent manners. J. Biol. Chem. **282**, 15471–15475
33. Bradley, C., van der Meer, R., Roodi, N., Yan, H., Chandrasekharan, M.B., Sun, Z.W., Mernaugh, R.L. and Parl, F.F. (2007) Carcinogen-induced histone alteration in normal human mammary epithelial cells. Carcinogenesis **28**, 2184–2192
34. Schulte, J.H., Lim, S., Schramm, A., Friedrichs, N., Koster, J., Versteeg, R., Ora, I., Pajtler, K., Klein-Hitpass, L., Kuhfittig-Kulle, S. et al. (2009) Lysine-specific demethylase 1 is strongly expressed in poorly differentiated neuroblastoma: implications for therapy. Cancer Res. **69**, 2065–2071

35. Wang, Y., Hacker, A., Murray-Stewart, T., Frydman, B., Valasinas, A., Fraser, A.V., Woster, P.M. and Casero, Jr, R.A. (2005) Properties of recombinant human N^1-acetylpolyamine oxidase (hPAO): potential role in determining drug sensitivity. Cancer Chemother. Pharmacol. **56**, 83–90
36. Huang, Y., Greene, E., Murray-Stewart, T., Goodwin, A., Baylin, S., Woster, P. and Casero, Jr, R. (2007) Inhibition of lysine-specific demethylase 1 by polyamine analogues results in reexpression of aberrantly silenced genes. Proc. Natl. Acad. Sci. U.S.A. **104**, 8023–8028
37. Stavropoulos, P. and Hoelz, A. (2007) Lysine-specific demethylase 1 as a potential therapeutic target. Expert Opin. Ther. Targets **11**, 809–820
38. Huang, Y., Murray Stewart, T., Wu, Y., Marton, L., Woster, P. and Casero, R. (2009) Novel oligoamine/polyamine analogues inhibit lysine-specific demethylase 1 (LSD1), induce re-expression of epigenetically silenced genes, and inhibit the growth of established human tumors *in vivo*. AACR 100th Annual Meeting, Denver, CO, U.S.A., 18–22 April 2009, Abstract LB-173
39. Murray-Stewart, T., Huang, Y., Woster, P. and Casero, R. (2008) Polyamine analogue inhibition of lysine-specific demethylase 1 in human acute myeloid leukemia cell lines. Proc. Am. Assoc. Cancer Res. **49**, 2605
40. McGarvey, K., Fahrner, J., Greene, E., Martens, J., Jenuwein, T. and Baylin, S. (2006) Silenced tumor suppressor genes reactivated by DNA demethylation do not return to a fully euchromatic chromatin state. Cancer Res. **66**, 3541–3549
41. Pruitt, K., Zinn, R.L., Ohm, J.E., McGarvey, K.M., Kang, S.H., Watkins, D.N., Herman, J.G. and Baylin, S.B. (2006) Inhibition of SIRT1 reactivates silenced cancer genes without loss of promoter DNA hypermethylation. PLoS Genet. **2**, e40
42. Cameron, E., Bachman, K., Myohanen, S., Herman, J. and Baylin, S. (1999) Synergy of demethylation and histone deacetylase inhibition in the re-expression of genes silenced in cancer. Nat. Genet. **21**, 103–107
43. Gore, S.D., Baylin, S., Sugar, E., Carraway, H., Miller, C.B., Carducci, M., Grever, M., Galm, O., Dauses, T., Karp, J.E. et al. (2006) Combined DNA methyltransferase and histone deacetylase inhibition in the treatment of myeloid neoplasms. Cancer Res. **66**, 6361–6369

Essays Biochem. (2009) **46**, 111–124; doi:10.1042/BSE0460008

8

Polyamines as mediators of APC-dependent intestinal carcinogenesis and cancer chemoprevention

Nathaniel S. Rial*†, Frank L. Meyskens, Jr‡ and Eugene W. Gerner*§[1]

**Department of Internal Medicine, The University of Arizona, 1501 North Campbell Avenue, Tucson, AZ 85724 U.S.A., †Cancer Biology, Arizona Cancer Center, 1515 North Campbell Avenue, Tucson, AZ 85724, U.S.A., ‡Chao Family Comprehensive Cancer Center and Departments of Medicine and Epidemiology, University of California, Irvine, CA 92868 U.S.A., and §The University of Arizona, Department of Cell Biology and Anatomy and Gastrointestinal Cancer Program, Arizona Cancer Center, 1515 North Campbell Avenue, Tucson, AZ 85724, U.S.A.*

Abstract

Combination chemoprevention for cancer was proposed a quarter of a century ago, but has not been implemented in standard medical practice owing to limited efficacy and toxicity. Recent trials have targeted inflammation and polyamine biosynthesis, both of which are increased in carcinogenesis. Preclinical studies have demonstrated that DFMO (difluoromethylornithine), an irreversible inhibitor of ODC (ornithine decarboxylase) which is the first enzyme in polyamine biosynthesis, combined with NSAIDs (non-steroidal

[1]*To whom correspondence should be addressed (email egerner@azcc.arizona.edu).*

anti-inflammatory drugs) suppresses colorectal carcinogenesis in murine models. The preclinical rationale for combination chemoprevention with DFMO and the NSAID sulindac, was strengthened by the observation that a SNP (single nucleotide polymorphism) in the *ODC* promoter was prognostic for adenoma recurrence in patients with prior sporadic colon polyps and predicted reduced risk of adenoma in those patients taking aspirin. Recent results from a phase III clinical trial showed a dramatic reduction in metachronous adenoma number, size and grade. Combination chemoprevention with DFMO and sulindac was not associated with any serious toxicity. A non-significant trend in subclinical ototoxicity was detected by quantitative audiology in a subset of patients identified by a genetic marker. These preclinical, translational and clinical data provide compelling evidence for the efficacy of combination chemoprevention. DFMO and sulindac is a rational strategy for the prevention of metachronous adenomas, especially in patients with significant risk for colorectal cancer. Toxicities from this combination may be limited to subsets of patients identified by either past medical history or clinical tests.

Introduction

Increased concentrations of polyamines are found in cancerous tissue [1]. Polyamines are influenced by factors such as import, export, biosynthesis and catabolism. ODC (ornithine decarboxylase) is a rationale target in several cancers, with the goal to decrease cellular polyamine pools. ODC inhibition may not be sufficient therapy in situations where polyamine synthesis is not the rate-limiting step regulating polyamine pool sizes.

The *ODC* gene is regulated by the Wnt signalling cascade [2]. Activated WNT signalling up-regulates MYC [3], a transcriptional activator of *ODC* [4]. The WNT cascade is silenced in the majority of adult intestinal tissues, but can be dysregulated through mutations. The *APC* (adenomatous polyposis coli) tumour suppressor gene is a component of the WNT cascade and has been identified as the germ-line mutation in *FAP* (familial adenomatous polyposis) [5]. Over 80% of sporadic colorectal cancers also carry mutations in *APC*. *APC* mutations and WNT signalling lead to up-regulation of MYC activity and increased expression of ODC, and increased polyamine pools. Other genetic influences of CRC (colorectal cancer) are mutant *K-RAS*, a proto-oncogene that also activates ODC.

Colorectal carcinogenesis is also influenced by high-fat diet and sedentary lifestyle. Long-term use of aspirin and the reduction of CRC mortality led to the COX (cyclo-oxygenase)-based hypothesis [6]. These environmental risk factors influence an individual's genetic susceptibility and produce a risk profile for cancer. A mechanism for CRC involves mutated-APC and K-RAS in combination with environmental risk factors that elevate polyamines and inflammation. A rationale strategy for cancer chemoprevention would be the combination of inhibitors of polyamine biosynthesis and inflammation.

Pre-clinical data

Cancer burden

In the United States, cancer is the leading cause of death in people under the age of 85 years [7]. There were an estimated 1437180 new cases and 565650 deaths attributed to cancer in 2008, whereas 10.8 million Americans are living with a history of cancer. Cancer therapeutics are most effective in the early stages of disease, but are less effective in treating advanced cancers. This point underscores the need for prevention, early detection and effective treatment.

The vast majority of US cancer incidence and mortality are epithelial-derived cancers of the lung, colon, breast and prostate. Epithelia provide a protective barrier while also importing nutrients and exporting waste, especially the colon. The ACS (American Cancer Society) estimated that there were 148810 new cases of colorectal cancer in 2008, with a mortality of 49960 deaths.

APC: the genetic risk factor for CRC

Carcinogenesis occurs by inactivation of a tumour suppressor gene or activation of an oncogene. Inherited loss of a single allele in the *APC* gene increases an individual's risk of a hereditary carcinogenesis, known as FAP [5]. FAP is characterized by an increase in adenomatous polyps of the colon with advancing age. The risk is malignant transformation of the polyps and treatment is colectomy. Greater than 80% of sporadic CRCs also have mutations in the *APC* gene. Shown in Figure 1 is a representation of the WNT pathway in normal tissue and in carcinogenesis.

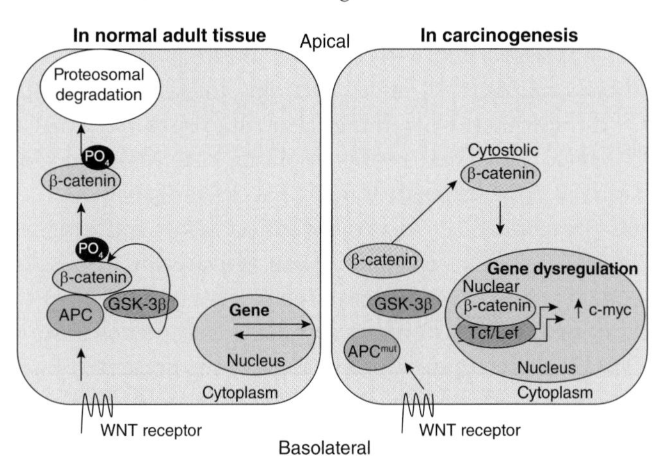

Figure 1. The WNT pathway in normal and carcinogenic cells

The WNT pathway in a normal, adult, colonic epithelial cell is depicted on the left-hand side of the Figure. On the basolateral portion of the cell the WNT receptor is activated and transmits the signal intracellularly to the APC, GSK-3β and β-catenin complex. GSK-3β phosphorylates β-catenin, marking it for proteosomal degradation. The WNT pathway in carcinogenesis is depicted on the right-hand side of the Figure. The WNT receptor may be activated and transmission of the signal intracellularly may occur, but inactivation of β-catenin by GSK-3β does not. Cytoplasmic accumulation of β-catenin leads to its nuclear translocation and binding with its cognate partner TCF/LEF. This heterodimerization regulates genes through transcription, notably of MYC.

The APC protein interacts with GSK-3β (glycogen synthase kinase-3β) and β-catenin. This interaction allows GSK-3β to phosphorylate β-catenin and marks it for proteosomal degradation. In carcinogenic tissue, GSK-3β and β-catenin are disrupted by mutant-APC. As a result GSK-3β can no longer phosphorylate β-catenin. β-Catenin accumulates in the cytoplasm and then translocates to the nucleus forming a heterodimer with its cognate binding partner, TCF/LEF (T-cell factor/lymphoid-enhancing factor). The heterodimer binds to promoter regions of genes and alters expression [2]. Dysregulation of WNT signalling leads to changes in *c-MYC* gene expression. MYC activation occurs in neuroblastoma, breast and prostate cancers, but its dysregulation in CRC is unique given mutations in the upstream *APC* gene and rapid cellular turnover.

In APC-dependent carcinogenesis c-MYC activation affects transcription of *ODC* through binding regions in the promoter. An SNP (single nucleotide polymorphism) occurs in *ODC* between two MYC-binding regions. Differential repression of *ODC* occurs by MAD1 binding at the SNP sequence to the A-allele, not the G-allele. In a population-based study of humans with prior colonic adenomas, aspirin use was associated with a 90% reduction in relative risk for development of metachronous adenomas among individuals homozygous for the minor *ODC* A-allele compared with the non-aspirin users homozygous for the major *ODC* G-allele [8]. This ODC polymorphism appears to be a genetic marker for CRC risk. Our group has sought to test features of the hypothesis depicted in Figure 1. We have developed a transgenic mouse expressing a mutant *Apc* and conditional deletion of the *c-Myc* alleles in the intestinal and colonic mucosa. Conditional suppression of Myc in the intestinal tract was associated with reduced intestinal tumour number, compared with these same *Apc*$^{\mathrm{Min}/+}$ mice expressing intestinal Myc [9]. Treatment with DFMO (difluoromethylornithine) also suppresses intestinal tumorigenesis in *Apc*$^{\mathrm{Min}/+}$ mice [10]. These results implicate both Myc and Odc as downstream mediators of APC-dependent intestinal carcinogenesis.

The MYC protein can act as a transcription factor activator when bound to MAX. The MYC–MAX heterodimer can activate the expression of genes through binding on consensus sequences (enhancer box sequences or E-boxes), as well as recruiting HATs (histone acetyltransferases). In contrast, when MAX is bound to MAD1, transcription is repressed. In the presence of normal APC, C-MYC was suppressed whereas MAD1 was elevated. In the presence of mutant-APC, C-MYC was elevated whereas MAD1 was suppressed [2]. This interaction is shown in Figure 2.

Mutant-*APC* led to elevations in c-MYC that increased expression of ODC [2]. The results indicated that ODC was a modifier of APC-dependent signalling in models of CRC. Mouse models were employed to determine whether mutant-APC led to elevated polyamines *in vivo*. Murine models recapitulate this finding, with elevated polyamines found in the small intestine of *Apc*$^{\mathrm{Min}/+}$ mice [10].

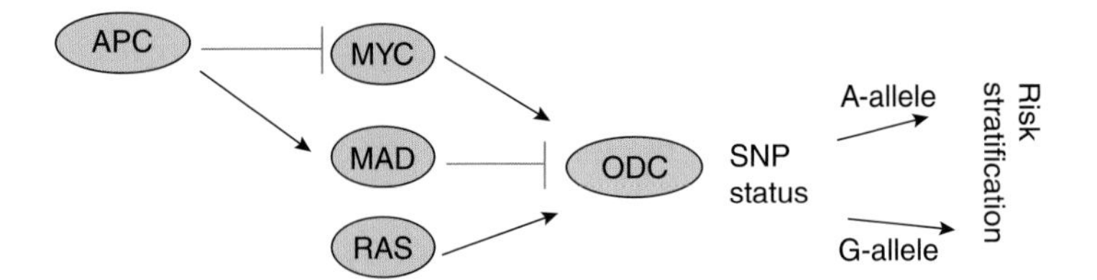

Figure 2. ODC regulation occurs through both positive and negative mechanisms
ODC is suppressed by MAD. Transcriptional activation of ODC can occur through MYC, RAS or both. Upstream activation of c-myc may occur via mutant APC. Pharmacogenetic manipulation of ODC occurs through identification of SNP status. The combination of these factors may increase individual risk stratification.

Polyamine-dependent regulation and colon carcinogenesis

Arginine is a component of the urea cycle which converts nitrogenous waste for excretion [11]. Elevated dietary arginine can also increase polyamine levels through its conversion into ornithine in the urea cycle. $Apc^{Min/+}$ mice fed a diet of elevated arginine had an elevated tumour burden [12]. In patients with CRC, increased meat consumption was a surrogate for arginine and was associated with decreased overall survival [13]. Polyamines are exported, as depicted in Figure 3, via a mechanism which involves an arginine transporter [14].

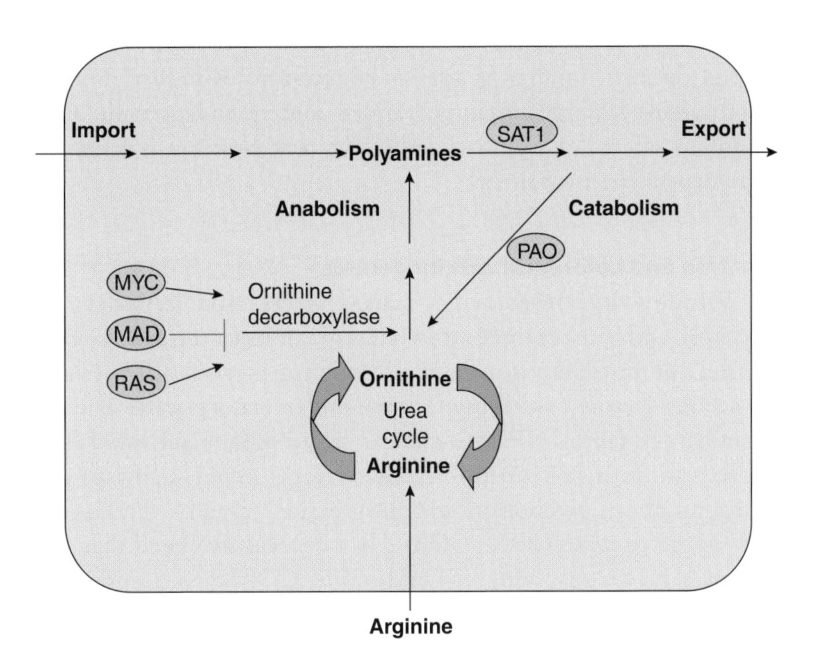

Figure 3. Polyamine transport is shown schematically via import, export, anabolism and catabolism
Arginine is imported into the cell and then converted into ornithine which contributes to the polyamine pools. Polyamine pools may be further increased by import or decreased by catabolism and export.

K-RAS mutation is another significant risk factor for CRC. The RAS-family of proteins is a mediator of extracellular signals through the cytoplasm and eventually into the nucleus; the effect of which is to alter gene expression and enhance proliferation. In cell culture models, mutant *K-RAS* increases ODC transcription and decreases transcription of SAT1 (spermidine/spermine acetyltransferase; also referred to as SSAT in other chapters in this volume), an enzyme important in the catabolism of polyamines. Mutant *K-RAS* increases polyamine biosynthesis via ODC activation and decreases acetylation of polyamines.

Polyamine pool limitation

One strategy to inhibit polyamine levels is to decrease biosynthesis. Selective inhibition of ODC by the suicide-inhibitor, α-DFMO, was developed at the Merrell Dow Research Center in Strasbourg, France [11]. Although α-DFMO showed promise in cell culture models, compensatory polyamine import limited the success of DFMO in early murine models.

SAT1 is an important factor in polyamine export. As shown in Figure 3, SAT1 can acetylate both spermidine and spermine, targeting them for export. *SAT1* can be induced by NSAIDs (non-steroidal anti-inflammatory drugs), including aspirin, sulindac, ibuprofen and indomethacin [15]. *SAT1* induction can lead to apoptosis in CRC cell lines [16]. Sulindac induced *Sat1*, decreased intestinal levels of monoacetylspermidine, spermidine and spermine, and reduced tumour number in the small intestine of mouse models [17]. Dietary putrescine restored tissue polyamine content and partially abrogated the antitumour effects of sulindac, indicating that sulindac was acting via a polyamine-dependent mechanism.

Inflammation and colorectal carcinogenesis

In 1863 Virchow hypothesized a causal interaction between chronic inflammation and cancer mediated via the tumour microenvironment [18]. Within the microenvironment of carcinogenesis both intrinsic and extrinsic cellular factors contribute to a pro-inflammatory state. The intrinsic pro-inflammatory factor NF-κB (nuclear factor κB) is activated by many signals. Activation of NF-κB up-regulates target genes facilitating cancer growth by initiation, promotion and progression. One of the target genes of NF-κB is the enzyme COX. *COX-1* is a constitutive gene that mediates homoeostatic functions, whereas inducible *COX-2* is associated with inflammation. PGE_2 (prostaglandin E_2) is downstream of COX-2 and is associated with tumorigenesis (Figure 4). In cell culture models the NSAID sulindac decreased both COX-2- and PGE_2 synthase-mediated inflammation. While NSAIDs block COX-2, they also inhibit production of NO (nitric oxide) via inhibition of NOS-2 (inducible nitric oxide synthase). NO is a known activator of inflammation. As shown in Figure 4, arginine can be converted into NO, leading to increased inflammation.

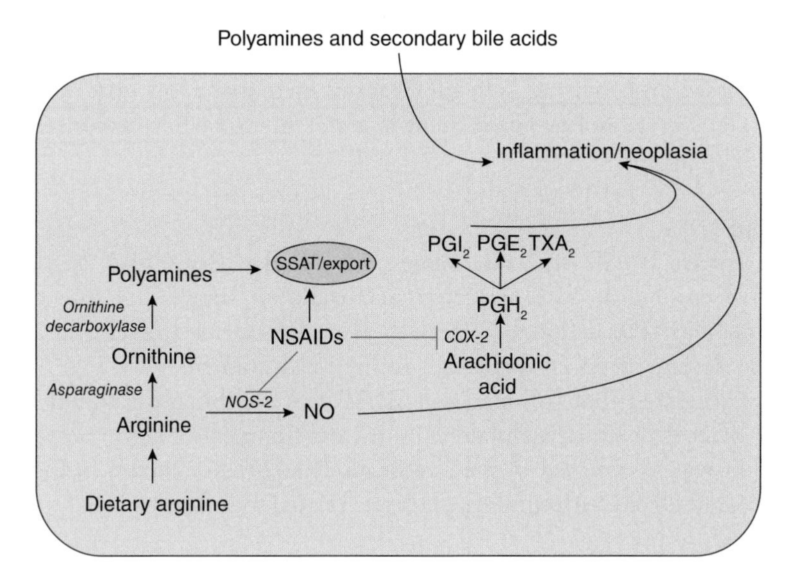

Figure 4. Inflammation within a colonic epithelial cell may occur through multiple mechanisms

Imported dietary arginine may either be converted into NO or processed in the polyamine pathway. NSAIDs can disrupt these and other pathways via inhibition of NOS-2, up-regulation of polyamine export or inhibition of COX-2. External sources of polyamines and bile acids may also contribute to inflammation.

Clinical data with DFMO

Cancer therapeutic experience

DFMO has been evaluated as a cancer therapeutic agent. It was not especially active as a single agent and its use was associated with ototoxicity at high doses [11,19]. Based on mouse model studies, DFMO was subsequently evaluated as a potential cancer chemopreventive agent [11].

Clinical trials of DFMO for colon cancer prevention

Early studies have confirmed that ODC and polyamine contents were elevated in human colon cancer tissue compared with adjacent normal colorectal mucosa [20]. Measurements of colorectal tissue polyamine contents were subsequently validated as measures of DFMO effects in patients. Validation of these markers allowed for the conduct of clinical trials to assess efficacy of DFMO dose, oral dose delivery and frequency of dosing. Consequently, we were able to conduct a dose de-escalation trial to determine the lowest DFMO dose capable of suppressing colorectal polyamine contents while minimizing toxicities, including ototoxicity [21]. A subsequent trial built on these findings evaluated three oral doses of DFMO given daily for 1 year. Patients were randomized to a control or treatment group with three separate doses of DFMO: 0.075, 0.2 and 0.4 g/m^2 per day [22]. The end of trial analysis showed that the 0.2 g/m^2 per day dose had similar biological effects compared with the 0.4 g/m^2 per day dose, with a decreased report of toxic side effects and decreased drop-out rate. The implications were that a low dose of DFMO inhibited colorectal polyamine content in rectal mucosa while demonstrating a safe toxicity profile.

Sulindac

While a phase IIb/III trial was being considered with DFMO, strong and extensive epidemiological evidence accumulated suggesting that aspirin and other NSAIDs might be effective as colon cancer prevention agents. Sulindac decreased polyp formation in high-risk patients with FAP [23]. It was hypothesized that sulindac and DFMO in combination would have a greater effect than either agent alone on the development of colon polyps, and this point was established in preclinical models [24]. As shown in Figure 5, NSAIDs and DFMO affect multiple targets related to inflammation.

Phase IIb and III DFMO and sulindac trials in non-cancerous patients

Cell culture models indicated that DFMO and sulindac could act at least additively to suppress growth and cell survival [24]. With mounting cell culture, murine model, clinical and epidemiological data, a prospective, randomized, placebo-controlled, phase IIb trial, with the combination of DFMO and sulindac, for 3 years was initiated in patients with prior sporadic

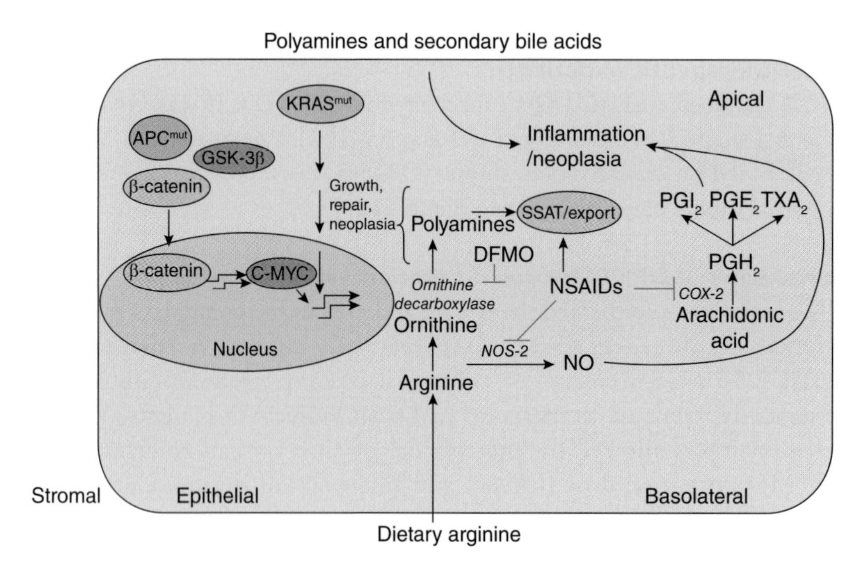

Figure 5. The interaction among mutant-APC, polyamines and inflammation is depicted in a colonic epithelial cell.
Mutant-APC and activated K-RAS lead to increased MYC production and up-regulate ODC which increases the polyamine concentration. Elevated dietary arginine can also contribute to increased polyamine pools. NSAIDs and DFMO can inhibit multiple targets in both the polyamine and inflammatory pathways.

colon polyps. The study used 0.2 g of DFMO/m^2 per day and was converted from a liquid oral dose into an oral pill form. A dose of 500 mg/day closely approximated the 0.2 g/m2 per day liquid form. The study also used 150 mg of sulindac, 50% of the conventional dose [23]. In addition to the need for efficacy, a major intent was to evaluate potential toxicity. The phase IIb trial of DFMO and sulindac, with biochemical markers as primary endpoints, was subsequently modified and converted into a phase III trial with metachronous adenomas as the primary endpoint of the study.

Baseline and serial audiological tests were performed to assess potential long-term ototoxicity. Participants with greater than 20 dB uncorrectable hearing loss above the age-adjusted norms were ineligible for the phase III trial. The eligible patients were randomized to receive placebo or 500 mg of DFMO plus 150 mg of sulindac. Patients were stratified according to two parameters: the seven clinical sites and low-dose aspirin use (either 81 mg/day or less than 325 mg twice weekly). Safety evaluations of the patients occurred after the 1 month run-in as well as at 3, 6 and 9 months, and every 6 months for the remainder of the phase III trial. Evaluations included physical examination and laboratory evaluations. Pure-tone audiograms were performed at 0, 18 and 36 months. Adenomas removed during any part of the phase III trial were submitted to a central pathology facility with standardized diagnostic objectives. The colorectal polyps were counted, measured and graded according to predetermined criteria. Safety analysis included investigator-reported adverse events and were coded according to the COSTART (Coding Symbols for Thesaurus of Adverse Reaction Terms) Body System [25]. An independent DSMB (Data and Safety Monitoring Board) reviewed the safety and efficacy of the data twice yearly. The investigators were blinded to the results of the DSMB's findings throughout the phase IIb and phase III trials. A pre-specified early stopping point was made based on potential efficacious or futile results. Interim analysis was planned at approx. 60% and 80% of the total accrual of patient information.

Based on the primary and secondary endpoints of the phase III trial, as well as oversight by the DSMB, the blind study was broken at the second interim analysis. The phase III trial intervention of combination DFMO and sulindac was found to be significantly effective. The combination of 500 mg of DFMO and 150 mg of sulindac decreased the number and severity of adenoma recurrence without significant toxicity [25]. The results are presented in Table 1 and summarized here. There was a 70% reduction in metachronous adenomas for those in the treatment arm. In addition, there was a 92% reduction in advanced adenomas and a 95% decrease in the recurrence of multiple adenomas in patients in the treatment arm compared with those in the placebo arm. Sporn and Hong [26] wrote an accompanying editorial with this publication and stated that "the clinical results represent a landmark advance to [reduce the number] of cancer deaths".

Table 1. Adenoma recurrence as a function of intervention

The adenomas and pathologies are shown on the left-hand side, while intervention number and type are shown along the top. The risk ratio for development of adenomatous lesions based on intervention strategy is quantified and statistical significance is shown. Data taken from [25].

Pathology	Placebo (n = 129)	DFMO+sulindac (n = 138)	Risk ratio	95% Confidence interval	*P*
Any adenoma	53	17	0.30	0.18–0.49	<0.001
Advanced adenoma	11	1	0.085	0.011–0.65	0.001
Large (≥1 cm) advanced adenoma	9	1	0.10	0.013–0.81	0.004
Multiple adenomas	17	1	0.055	0.0074–0.41	<0.001

Toxicity

An adjusted, non-significant, mean decrease in hearing threshold of 1.08 dB was detected in the treatment arm compared with placebo [27]. There was no difference in the clinical audiotoxicity between the two arms. Cardiotoxicity was also evaluated in this study. Patients were stratified into low-, moderate- and high-CV (cardiovascular) risk factors. Among high-risk patients, the number of CV events was higher in the treatment than the placebo arm. Excluding the high-risk CV patients, the numbers of CV events were similar between treatment and placebo arms [28].

These clinical results from a phase III trial proved to be consistent with preclinical studies in mouse models. Combination DFMO and sulindac was effective in reducing tumour number by more than 80% when compared with the untreated controls in the $Apc^{Min/+}$ mouse model ($P < 0.0001$) [29]. Combination sulindac–DFMO was effective in reducing the number of high-grade adenomas when compared with the sulindac alone ($P = 0.003$). The clinical implications are twofold: (i) first DFMO is effective in the reduction of adenomatous polyps and (ii) secondly, the combination of DFMO and sulindac can further reduce the risk of CRC through the reduction in the number of high-grade intestinal adenomas. These high-grade adenomas are those lesions most likely to progress to colon cancer.

Phase III DFMO and sulindac in patients with cancer

Although the phase III trial with DFMO and sulindac provided strong evidence for preventing disease recurrence in patients with prior adenomas, the chemoprevention has not been evaluated in higher risk populations. Even after surgical resection and optimal treatment with chemotherapy (when indicated), stage I–III colon cancer patients remain at considerable risk for distant recurrence, secondary colonic tumour formation and subsequent mortality. A phase III trial of these key compounds among surgically resected colon cancer patients is in the planning stages.

Future

Clinical practice

Approx. 30 million patients over the age of 50 years will develop adenomatous polyps each year in the United States. A proportionally similar number will develop these lesions in Western Europe. Approx. 10% will progress to advanced polyps or frank cancer. Genetically at-risk patients along with those with a history of CRC [sporadic, FAP, HNPCC (hereditary nonpolyposis colorectal cancer) or prior CRC] are the target population for chemoprevention. Current techniques for CRC screening include endoscopy at regular intervals with polyps removed. There is a significant risk reduction for CRC through endoscopy and polypectomy, but barriers of cost, preparation and the procedure translates into approx. 50% screening of eligible populations.

Endoscopy is also a poor screening method for right-sided cancers, as well as flat or depressed lesions. Further complicating the role of endoscopy is the frequent overuse by patients with previously diagnosed disease. The resultant gaps in screening with overuse by others necessitate another disease prevention strategy. Chemoprevention clearly has the potential to reduce disease burden.

The results from the phase III trials potentially affect the surveillance of higher risk populations. As previously stated, approx. 50% of CRC patients with Stage I, II or III disease recur. Patients' and physicians' anxieties may decrease with combination therapy and lessen the overutilization of endoscopic procedures in patients with a moderate risk of colon cancer. Combination chemoprevention may also have applications in high-risk populations such as those with FAP. The positive results of this combination in the *Apc* $^{Min/+}$ model of FAP may lead to trials in this high-risk group. Positive results of such trials could increase the time to surgery in patients with FAP.

Dietary sources of polyamines

A database has been developed to assess dietary polyamine content [30]. It is hypothesized that a reduction in total-body polyamine pools may reduce carcinogenesis. Table 2 presents some of the foods for which the polyamine content has been determined. The intention for the database is to quantify dietary polyamines and qualify them as a risk factor for carcinogenesis.

Conclusions

Epithelia provide a protective barrier from the external environment. It is a mediator of transport. In the process, epithelia are exposed to harmful substances and harbour genetic mutations. In this context, cells are transformed from normal to neoplastic and the carcinogenic process is initiated. For the past

Table 2. Polyamine content in food

The three polyamines: putrescine, spermidine and spermine are listed. The amount of polyamines, in nmol/day, is given for multiple food items. Data taken from [30].

Polyamine	Food item	Amount (nmol/day)
Putrescine	Orange juice and grapefruit juice	44441
	Oranges, grapefruit and tangerines (excluding juice)	17613
	Fresh tomatoes	10042
	Bananas	7344
	Beer (all types)	6374
Spermidine	Green peas	3283
	Cheese, such as American and cheddar	3124
	Lasagne and pasta with meat sauce	2900
	Potatoes (boiled, baked and mashed)	2388
	Burritos, tacos, tostadas and quesadillas	1890
Spermine	Ground meat	2186
	Lunch meats (e.g. ham, turkey, bologna and salami)	1977
	Green peas	1905
	Lasagne and pasta with meat sauce	1443
	Peanut butter, peanuts and other nuts and seeds	1237

several decades, collaborative efforts of scientific research have led to a new paradigm in cancer management: chemoprevention. Chemoprevention is an approach that is not applicable to everyone. It should target people with elevated risks of cancer. The caveat for chemoprevention is that target populations are still relatively healthy compared with patients who currently have cancer. Therefore chemoprevention must have a clear benefit that exceeds the risk of treatment. The development of DFMO and sulindac underscore this sentiment.

Summary

- *Polyamine pools are elevated in cancerous tissue.*
- *ODC is a committed step, converting ornithine into putrescine.*
- *DFMO is an irreversible, competitive inhibitor of ODC.*
- *In CRC cell culture models, DFMO decreased polyamine levels.*
- *Murine models of CRC with mutant-Apc have increased Myc and Odc activation.*
- *DFMO alone had limited efficacy in therapeutic clinical trials.*
- *NSAIDs decrease COX activity.*
- *NSAIDs can activate polyamine export pathways.*
- *Polyamine pools are dynamically regulated by anabolism, catabolism, import and export.*

- *Low-dose DFMO was evaluated as a chemopreventative agent for CRC by biomarker assays, efficacy, safety, toxicity and dose de-escalation before large clinical trials.*
- *Combination DFMO and sulindac was evaluated in a large, randomized, prospective, multi-centre, double-blind clinical trial.*
- *Combination DFMO and sulindac was effective in reducing adenoma recurrence by number, size and grade.*
- *Ototoxicity was measured in the treatment arm, but without clinical significance.*
- *Patients with baseline high-CV risk had a higher number of CV events.*

References

1. Russell, D. and Snyder, S.H. (1968) Amine synthesis in rapidly growing tissues: ornithine decarboxylase activity in regenerating rat liver, chick embryo, and various tumors. Proc. Natl. Acad. Sci. U.S.A. **60**, 1420–1427
2. Fultz, K.E. and Gerner, E.W. (2002) APC-dependent regulation of ornithine decarboxylase in human colon tumor cells. Mol. Carcinog. **34**, 10–18
3. He, T.C., Sparks, A.B., Rago, C., Hermeking, H., Zawel, L., da Costa, L.T., Morin, P.J., Vogelstein, B. and Kinzler, K.W. (1998) Identification of c-MYC as a target of the APC pathway. Science **281**, 1509–1512
4. Bello-Fernandez, C., Packham, G. and Cleveland, J.L. (1993) The ornithine decarboxylase gene is a transcriptional target of c-Myc. Proc. Natl. Acad. Sci. U.S.A. **90**, 7804–7808
5. Groden, J., Thliveris, A., Samowitz, W., Carlson, M., Gelbert, L., Albertsen, H., Joslyn, G., Stevens, J., Spirio, L., Robertson, M. et al. (1991) Identification and characterization of the familial adenomatous polyposis coli gene. Cell **66**, 589–600
6. Thun, M.J., Namboodiri, M.M. and Heath, Jr, C.W. (1991) Aspirin use and reduced risk of fatal colon cancer. N. Engl. J. Med. **325**, 1593–1596
7. Jemal, A., Siegel, R., Ward, E., Murray, T., Xu, J. and Thun, M.J. (2007) Cancer statistics, 2007. CA Cancer J. Clin. **57**, 43–66
8. Martinez, M.E., O'Brien, T.G., Fultz, K.E., Babbar, N., Yerushalmi, H., Qu, N., Guo, Y., Boorman, D., Einspahr, J., Alberts, D.S. and Gerner, E.W. (2003) Pronounced reduction in adenoma recurrence associated with aspirin use and a polymorphism in the ornithine decarboxylase gene. Proc. Natl. Acad. Sci. U.S.A. **100**, 7859–7864
9. Ignatenko, N.A., Holubec, H., Besselsen, D.G., Blohm-Mangone, K.A., Padilla-Torres, J.L., Nagle, R.B., de Alboranc, I.M., Guillen, R.J. and Gerner, E.W. (2006) Role of c-Myc in intestinal tumorigenesis of the $Apc^{Min/+}$ mouse. Cancer Biol. Ther. **5**, 1658–1664
10. Erdman, S.H., Ignatenko, N.A., Powell, M.B., Blohm-Mangone, K.A., Holubec, H., Guillen-Rodriguez, J.M. and Gerner, E.W. (1999) APC-dependent changes in expression of genes influencing polyamine metabolism, and consequences for gastrointestinal carcinogenesis, in the Min mouse. Carcinogenesis **20**, 1709–1713
11. Gerner, E.W. and Meyskens, Jr, F.L. (2004) Polyamines and cancer: old molecules, new understanding. Nat. Rev. Cancer **4**, 781–792
12. Yerushalmi, H.F., Besselsen, D.G., Ignatenko, N.A., Blohm-Mangone, K.A., Padilla-Torres, J.L., Stringer, D.E., Guillen, J.M., Holubec, H., Payne, C.M. and Gerner, E.W. (2006) Role of polyamines in arginine-dependent colon carcinogenesis in $Apc^{Min/+}$ mice. Mol. Carcinog. **45**, 764–773
13. Zell, J.A., Ignatenko, N.A., Yerushalmi, H.F., Ziogas, A., Besselsen, D.G., Gerner, E.W. and Anton-Culver, H. (2007) Risk and risk reduction involving arginine intake and meat consumption in colorectal tumorigenesis and survival. Int. J. Cancer **120**, 459–468

14. Uemura, T., Yerushalmi, H.F., Tsaprailis, G., Stringer, D.E., Pastorian, K.E., Hawel, III, L., Byus, C.V. and Gerner, E.W. (2008) Identification and characterization of a diamine exporter in colon epithelial cells. J. Biol. Chem. **283**, 26428–26435
15. Babbar, N., Gerner, E.W. and Casero, Jr, R.A. (2006) Induction of spermidine/spermine N1-acetyltransferase (SSAT) by aspirin in Caco-2 colon cancer cells. Biochem. J. **394**, 317–324
16. Babbar, N., Ignatenko, N.A., Casero, Jr, R.A. and Gerner, E.W. (2003) Cyclooxygenase- independent induction of apoptosis by sulindac sulfone is mediated by polyamines in colon cancer. J. Biol. Chem. **278**, 47762–47775
17. Ignatenko, N.A., Besselsen, D.G., Roy, U.K., Stringer, D.E., Blohm-Mangone, K.A., Padilla-Torres, J.L., Guillen, R.J. and Gerner, E.W. (2006) Dietary putrescine reduces the intestinal anticarcinogenic activity of sulindac in a murine model of familial adenomatous polyposis. Nutr. Cancer **56**, 172–181
18. Coussens, L.M. and Werb, Z. (2002) Inflammation and cancer. Nature **420**, 860–867
19. Croghan, M.K., Aickin, M.G. and Meyskens, F.L. (1991) Dose-related α-difluoromethylornithine ototoxicity. Am. J. Clin. Oncol. **14**, 331–335
20. Hixson, L.J., Garewal, H.S., McGee, D.L., Sloan, D., Fennerty, M.B., Sampliner, R.E. and Gerner, E.W. (1993) Ornithine decarboxylase and polyamines in colorectal neoplasia and mucosa. Cancer Epidemiol. Biomarkers Prev. **2**, 369–374
21. Meyskens, Jr, F.L. Emerson, S.S., Pelot, D., Meshkinpour, H., Shassetz, L.R., Einspahr, J., Alberts, D.S. and Gerner, E.W. (1994) Dose de-escalation chemoprevention trial of α-difluoromethylornithine in patients with colon polyps. J. Natl. Cancer Inst. **86**, 1122–1130
22. Meyskens, Jr, F.L., Gerner, E.W., Emerson, S., Pelot, D., Durbin, T., Doyle, K. and Lagerberg, W. (1998) Effect of α-difluoromethylornithine on rectal mucosal levels of polyamines in a randomized, double-blinded trial for colon cancer prevention. J. Natl. Cancer Inst. **90**, 1212–1218
23. Giardiello, F.M., Offerhaus, J.A., Tersmette, A.C., Hylind, L.M., Krush, A.J., Brensinger, J.D., Booker, S.V. and Hamilton, S.R. (1996) Sulindac induced regression of colorectal adenomas in familial adenomatous polyposis: evaluation of predictive factors. Gut **38**, 578–581
24. Lawson, K.R., Ignatenko, N.A., Piazza, G.A., Cui, H. and Gerner, E.W. (2000) Influence of K-ras activation on the survival responses of Caco-2 cells to the chemopreventive agents sulindac and difluoromethylornithine. Cancer Epidemiol. Biomarkers Prev. **9**, 1155–1162
25. Meyskens, F.L., McLaren, C.E., Pelot, D., Fujikawa-Brooks, S., Carpenter, P.M., Hawk, E., Kelloff, G., Lawson, M.J., Kidao, J., McCracken, J. et al. (2008) Difluoromethylornithine plus sulindac for the prevention of sporadic colorectal adenomas: a randomized placebo-controlled, double-blind trial. Cancer Prev. Res. **1**, 9–11
26. Sporn, M.B. and Hong, W.K. (2008) Clinical prevention of recurrence of colorectal adenomas by the combination of difluoromethylornithine and sulindac: an important milestone. Cancer Prev. Res. **1**, 9–11
27. McClaren, C.E., Fujikawa-Brooks, S., Chen, W., Gillen, D.L., Pelot, D., Gerner, E.W. and Meyskens, F.L. (2008) Longitudinal assessment of air conduction audiograms in a phase III clinical trial of difluoromethylornithine and sulindac for prevention of sporadic colorectal adenomas. Cancer Prev. Res. **1**, 514–521
28. Zell, J.A., Pelot, D., Chen, W.P., McLaren, C.E., Gerner, E.W. and Meyskens, F.L. (2009) Risk of cardiovascular events in a randomized placebo-controlled, double-blind trial of difluoromethylornithine plus sulindac for the prevention of sporadic colorectal adenomas. Cancer Prev. Res. **2**, 209–212
29. Ignatenko, N.A., Besselsen, D.G., Stringer, D.E., Blohm-Mangone, K.A., Cui, H. and Gerner, E.W. (2008) Combination chemoprevention of intestinal carcinogenesis in a murine model of familial adenomatous polyposis. Nutr. Cancer **60** (Suppl. 1), 30–35
30. Zoumas-Morse, C., Rock, C.L., Quintana, E.L., Neuhouser, M.L., Gerner, E.W. and Meyskens, Jr, F.L. (2007) Development of a polyamine database for assessing dietary intake. J. Am. Diet Assoc. **107**, 1024–1027

Essays Biochem. (2009) **46**, 125–144; doi:10.1042/BSE0460009

9

Transgenic animals modelling polyamine metabolism-related diseases

Leena Alhonen[1], Anne Uimari, Marko Pietilä, Mervi T. Hyvönen, Eija Pirinen and Tuomo A. Keinänen

A.I. Virtanen Institute for Molecular Sciences, Biocenter Kuopio, University of Kuopio, P.O. Box 1627, FI-70211 Kuopio, Finland

Abstract

Cloning of genes related to polyamine metabolism has enabled the generation of genetically modified mice and rats overproducing or devoid of proteins encoded by these genes. Our first transgenic mice overexpressing *ODC* (ornithine decarboxylase) were generated in 1991 and, thereafter, most genes involved in polyamine metabolism have been used for overproduction of the respective proteins, either ubiquitously or in a tissue-specific fashion in transgenic animals. Phenotypic characterization of these animals has revealed a multitude of changes, many of which could not have been predicted based on the previous knowledge of the polyamine requirements and functions. Animals that overexpress the genes encoding the inducible key enzymes of biosynthesis and catabolism, ODC and SSAT (spermidine/spermine N^1-acetyltransferase) respectively, appear to possess the most pleiotropic phenotypes. Mice overexpressing *ODC* have particularly been used as cancer research models. Transgenic mice and rats with enhanced polyamine catabolism have revealed an association of rapidly depleted polyamine pools

[1]*To whom correspondence should be addressed (email Leena.Alhonen@uku.fi).*

and accelerated metabolic cycle with development of acute pancreatitis and a fatless phenotype respectively. The latter phenotype with improved glucose tolerance and insulin sensitivity is useful in uncovering the mechanisms that lead to the opposite phenotype in humans, Type 2 diabetes. Disruption of the *ODC* or *AdoMetDC* [AdoMet (*S*-adenosylmethionine) decarboxylase] gene is not compatible with mouse embryogenesis, whereas mice with a disrupted *SSAT* gene are viable and show no harmful phenotypic changes, except insulin resistance at a late age. Ultimately, the mice with genetically altered polyamine metabolism can be used to develop targeted means to treat human disease conditions that they relevantly model.

Introduction

The development of sophisticated gene-transfer techniques to manipulate the mammalian genome has enabled the study of the expression and function of individual genes in a whole-body context. With these techniques, the importance of any protein, structural or metabolism-related, can be subjected to evaluation in genetically modified animals. Based on the knowledge acquired, a considerable number of disease models mimicking various human pathologies have been generated.

The natural polyamines are critical for normal cellular growth and differentiation, and the tissue levels of individual polyamines are maintained optimal via complex regulatory mechanisms. Although these levels can be transiently manipulated by physiological stimuli or with inhibitors or inducers of polyamine metabolic enzymes, genetic manipulation is a method offering a versatile means to affect polyamine homoeostasis in different tissues and for an extended period, even throughout life. The animals with genetically altered polyamine synthesis or catabolism not only help to define the physiological importance of the polyamines in the whole body, but they also offer means to develop treatment modalities for pathophysiological conditions that result from distorted polyamine homoeostasis.

As polyamine synthesis is needed to maintain proliferation, and elevated polyamine levels are often associated with accelerated growth and cancer, it is not surprising that the first transgenic mice with genetically engineered polyamine metabolism were animals overexpressing the *ODC* (ornithine decarboxylase) gene encoding the key regulatory enzyme in polyamine biosynthesis. Depending on the extent and type of overexpression, somewhat controversial results have initially been reported. For instance, life-long ubiquitous overexpression of the *ODC* gene under its own promoter was, to our surprise, not associated with generally enhanced tumorigenesis in any tissue of the transgenic mice in comparison with wild-type littermates [1]. The main factor determining this phenotype is most likely the fact that putrescine was not converted further into spermidine and spermine in the tissues of these animals [2], indicating that powerful regulatory mechanisms prevent overaccumulation of the higher polyamines to potentially harmful levels. In contrast, hair-follicle-targeted expression of truncated *ODC* cDNA resulted in prominent dermal ODC activity, polyamine accumulation, hair loss and spontaneous formation of malignant tumours in skin [3]. Interestingly, the first visible characteristic of mice overexpressing ubiquitously the genomic construct encoding the key catabolic enzyme, SSAT (spermidine/spermine N^1-acetyltransferase), was their hairlessness, which was likewise related to accumulation of putrescine in the skin [4]. It is thus suggested that the overaccumulation of putrescine, either via *de novo* synthesis or via back conversion from spermidine and spermine, is a critical factor in the development of skin neoplasms.

In general, the altered phenotype of the transgenic mice may result from the perturbation of polyamine homoeostasis, from the impact of the accelerated polyamine metabolic cycle on other metabolic pathways, or from both.

Polyamine metabolism as target for genetic engineering

Figure 1 depicts the steps of the polyamine metabolic pathway and shows the enzymes that are the targets of overexpression or gene disruption. References for each transgenic line can be found in Table 1, albeit the list does not provide full coverage of characterization of the lines. The genes and cDNAs encoding the key enzymes of biosynthesis, the inducible ODC and AdoMetDC (*S*-adenosylmethionine decarboxylase), or catabolism, the inducible SSAT, have been used to generate both transgenic mice and rats. All of these genes have also been disrupted in knockout mice devoid of the respective enzyme proteins. Mice overexpressing Az (antizyme), a protein inhibitor of ODC, likewise serve as models possessing decreased ODC activity in targeted tissues. *SpmS* (spermine synthase)-overexpressing mice are phenotypically normal and viable [5], whereas knockout of this gene has

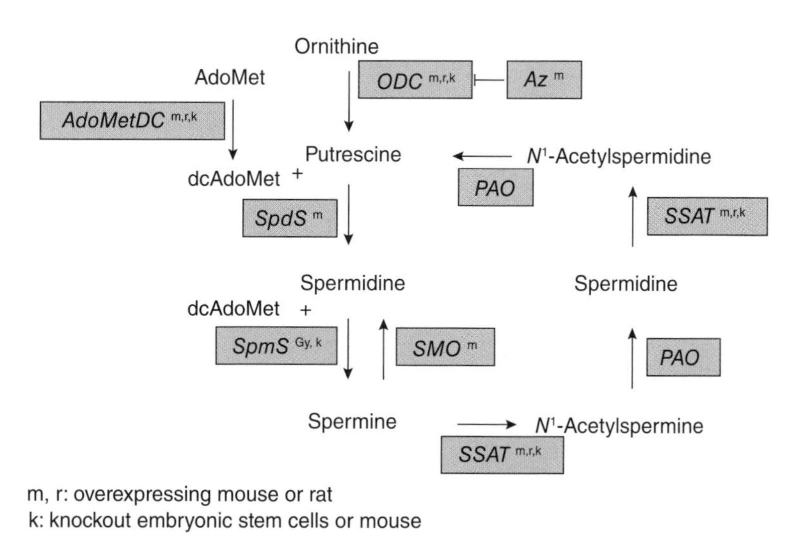

Figure 1. Polyamine metabolism
The targeted genes in genetically modified mice (m) or rats (r) are highlighted in blue.

Table 1. Genetically engineered rodent lines with altered polyamine metabolism

CMV-IE, cytomegalovirus immediate early; DENSPM, N^1,N^{11}-diethylnorspermine; MMTV-LTR, mouse mammary tumour virus long terminal repeat; NMBA, *N*-nitrosomethylbenzylamine.

Gene (coding sequence)	Promoter/type of expression	Animal species/ tissue-specificity of expression	Phenotypic changes (in comparison with wild-type mice)	References
(a) Overexpression of transgene				
ODC human genomic	Human genomic/constituve	Mouse/ubiquitous	Male infertility, enhanced papilloma formation	[37,40]
			Partial blockade of NMDA receptor	[17]
		Rat/ubiquitous	No marked changes	[41]
ODC human genomic	Metallothionein/inducible	Mouse/relatively broad	No marked changes	[42]
ODC mouse cDNA truncated	K5 or K6/constitutive	Mouse/skin	Hair loss, wrinkled skin, spontaneous	[3,43]
			Tumour formation, increased susceptibility to photocarcinogenesis	[44]
ODC mouse cDNA	MMTV-LTR/constitutive	Mouse/broad	Increased spontaneous tumour formation	[45]
ODC mouse cDNA truncated	α-Myosin heavy chain/constitutive	Mouse/heart	No marked changes	[46]
ODC mouse cDNA-oestrogen receptor	Involucrin/inducible	Mouse/suprabasal	Keratinocyte activation, stimulation of vascularization	[25]
AdoMetDC rat genomic	Genomic/constitutive	Mouse and rat/ubiquitous	No marked changes	[47]

AdoMetDC human cDNA truncated	α-Myosin heavy chain/constitutive	Mouse/heart	Increased isoprenaline-induced cardiac hypertrophy	[48]
Az rat cDNA mutated	K5 or K6/constitutive	Mouse/skin	Reduced papilloma formation	[8]
	K5/constitutive	Mouse/forestomach epithelium	Apoptosis, resistance to NMBA-induced carcinogenesis	[49]
Az rat cDNA mutated	α-Myosin heavy chain/constitutive	Mouse/heart	No changes in development of cardiac hypertrophy	[50]
SpdS human genomic	Genomic/constitutive	Mouse/ubiquitous	No marked changes	[51]
SpmS human cDNA	CMV-IE enhancer-chicken β-actin/ubiquitous	Mouse/ubiquitous	No marked changes	[5]
SSAT mouse genomic	Genomic/constitutive	Mouse/ubiquitous	Hair loss, wrinkled skin, shortened life-span, reduced fat mass, female infertility, partial resistance to skin carcinogenesis	[4,26]
			Central nervous system disturbances	[18,20]
			Reduced ATP and acetyl-CoA in white adipose tissue, enhanced insulin sensitivity, improved glucose tolerance	[31,32]
	Metallothionein/inducible	Mouse and rat/mainly pancreas and liver	Hair loss, wrinkled skin, lipoatrophy (mouse)	[24]

(Continued)

Table 1. *Continued*

Gene (coding sequence)	Promoter/type of expression	Animal species/ tissue-specificity of expression	Phenotypic changes (in comparison with wild-type mice)	References
			Development of acute pancreatitis (mouse and rat)	[28]
			Delayed liver regeneration (rat)	[52]
	CMV-tet/inducible	Mouse/ubiquitous	Development of pancreatitis	[53]
SSAT cDNA	K6/constitutive	Mouse/ubiquitous	Increased sensitivity to carcinogenesis	[54]
(b) Inactivation of gene (knockout)				
ODC$^{+/-}$			Decreased papilloma formation in carcinogenesis	[55]
ODC$^{-/-}$			Death during embryonic development	[56]
AdoMetDC$^{+/-}$			No changes	
AdoMetDC$^{-/-}$			Death during embryonic development	[57]
SSAT$^{-/-}$, *SSAT*$^{0/-}$			Insulin resistance upon aging	[33]

not been achieved in mice. However, embryonic stem cells with a disrupted *SpmS* gene have been established and studied [6]. Existence of a mutated mouse strain with an X-chromosomal deletion that affects *SpmS* and a phosphate-regulating gene Phex [7] compensates for the lack of gene-disrupted mice generated via targeted homologous recombination within the *SpmS* gene. The catabolic enzymes PAO (polyamine oxidase), catalysing the oxidation of acetylated spermidine and spermine, and SMO (spermine oxidase), catalysing the conversion of spermine directly into spermidine, appear to be the only enzymes immediately involved in polyamine metabolism, the overexpression of which in transgenic animals has not so far been reported.

Aspects and approaches in the production of rodent lines with genetically engineered polyamine metabolism

Methods of genetic modification

Mice or rats with transgene overexpression are usually produced via a conventional method of pronuclear injection of the transgene construct into one-cell embryos. The drawback of this method is the inability to control the site(s) of transgene integration and the copy number integrated. The random aberrant chromosomal environment may greatly influence the expression by either inducing or silencing it, and hence transgene copy number does not necessarily correlate with the level of expression. For gene inactivation, constructs with deleterious mutations disrupting the function of the genes are incorporated into embryonic stem cells via homologous recombination. These cells are used to produce chimaeric mice and eventually mice completely devoid of the functional gene. In this method, the site of manipulation is controlled and all the other genes remain intact. The lack of appropriate rat embryonic stem cells has unfortunately limited the use of this method to mice.

Transgene constructs

When we first started to use transgene techniques we chose genomic DNA, namely *ODC* and later *AdoMetDC*, *SpdS* (spermidine synthase) and *SSAT* genes under their own promoters, to work with. The idea was to have all of the regulatory elements in the constructs so that the expression of the transgenes would be regulated in a similar fashion to the expression of the endogenous genes. Such an approach would serve best the study of regulation of gene expression under different conditions *in vivo*. Others decided to use cDNAs of the same genes instead, with the obvious aim to achieve maximum overexpression, controlled by a strong and tissue-specific promoter and free of any potential silencer elements. The latter approach would serve the purpose of studying the consequences of robust overexpression in a given tissue. As an additional feature, inducibility of the promoter allows the follow-up of consequences of transgene induction at a desired time, whereas constitutive expression affects the animal throughout its life. As seen in Table 1, ubiquitous expression has been achieved with the genomic constructs, and targeted expression to selected tissues, such as skin, heart, pancreas and liver, has been achieved with the aid of tissue-specific promoters.

Genetic background of mice

Embryo production and quality is to some extent dependent on the mouse strain used. Depending on the preference of each transgene facility, transgenic lines coming from them may originally be generated and subsequently maintained in various mouse strains, which makes the direct comparison of transgenic lines difficult or questionable. It has also been noticed that phenotypic traits may be lost, or others acquired, upon backcrossing the mice to a different background from that originally used. C57Bl/6 is the most convenient mouse strain in terms of being the same as the background of the most knockout mice available. It should be pointed out here that, although we use embryos from a hybrid background (BalbC×DBA2) as recipients of transgenes and the initial characterization of transgenic mice has usually been carried out in the same background, the majority of phenotypic changes have been preserved upon later backcrossing the mice into a C57Bl/6 background. The importance of genetic background is also exemplified in the consequences of skin-targeted expression of Az: tumour incidence in response to two-stage carcinogenesis is decreased in a carcinogenesis resistant C57Bl/6J strain as compared with the sensitive DBA/2J strain, although both carry the same transgene in the same locus [8].

Phenotypic characterization

Despite the fact that success in generation of transgenic founders does not automatically guarantee expression of the transgenes and appearance of such a phenotype that is worth investigating, a great many transgenic mice seem to exhibit a noteworthy phenotype, often much more pleiotropic than expected at the time of construct design. When characterizing transgenic animals, it is initially necessary to compare two or more lines carrying the same transgene to exclude the potential effect of the integration site on the expression and the subsequent phenotype. In rare cases, transgene(s) may integrate within another gene causing a specific phenotype owing to inactivation of that gene and not exclusively to the expression of the transgene.

Table 1 lists most of the genetically modified rodents with altered polyamine metabolism that have been reported so far. The lines listed are singly transgenic lines, some of which show markedly changed phenotypes, whereas others carry no or very mild phenotypic changes in comparison with wild-type animals. The lack of a distinct phenotype in general may result from low transgene expression or some compensatory physiological responses. Because of the latter, no distinct phenotype may be observed, despite the transgene activity-driven alterations in polyamine homoeostasis. Comparison of phenotypes between individual mouse lines that overproduce the same enzyme may sometimes be confusing owing to controversial phenotypic characteristics. The phenotypic differences are obviously attributable to the differing regulatory elements in gene constructs, dissimilar levels of expression in a given tissue and the different genetic background of mice used. Fortunately, the observed phenotype in many transgenic mouse or rat lines may mimic some human diseases closely enough so that the animals can be used as models to study the pathogenetic processes of these diseases and the role of polyamines in them. The phenotypic changes in mouse and rat lines listed in Table 1 have been described in detail in extensive reviews [9–11].

Crossbreeding of genetically modified rodent lines together is a way to broaden the impact of genetic alterations and to study the modulatory effect of impaired polyamine homoeostasis on other models, especially those designed for carcinogenesis studies. For instance, doubly transgenic mice carrying both a K6-*ODC* transgene and a v-Ha-*ras* transgene developed spontaneous skin carcinomas, although either line alone did not [12], indicating the role for activated ODC and elevated polyamines in tumorigenesis. This conclusion is further supported by subsequent work showing that inhibition of ODC with DFMO (difluoromethylornithine) caused a regression of evolved tumours in K6-*ODC*/Ras mice [13]. Overexpression of Az in the skin of doubly transgenic Az/MEK [MAPK (mitogen-activated protein kinase)/ERK (extracellular-signal-regulated kinase) kinase] mice delayed tumorigenesis in comparison with MEK mice alone, leading to the same conclusion about the importance of polyamines in tumorigenesis [14]. Yet in another example, haplosufficiency for ODC delayed lymphomagenesis in immunoglobulin enhancer (Emu)-myc-transgenic mice crossed with heterozygously *ODC*-deficient mice [15]. In a recent study, deafness and sensitivity of Gy mice (carrying the *SpmS* deletion) to DFMO treatment could be reversed by crossing these mice with *SpmS*-overexpressing mice [16], emphasizing the importance of maintaining sufficient spermine levels in the inner ear.

Diseases associated with altered polyamine metabolism

Diseases associated with changes in polyamine levels

Central nervous system abnormalities

Our studies suggest an important beneficial role for putrescine in the brain of mice overexpressing either the *ODC* or *SSAT* gene under its own promoter. By using these transgenic animals, we have been able to show convincingly that putrescine accumulation elevates seizure threshold [17,18] and protects against ischaemia/reperfusion damage [19], but leads to impaired spatial learning [17,20]. These changes can be speculated to result from the role of putrescine as an antagonist of the NMDA (*N*-methyl-D-aspartate) receptor. Hence, the elevated putrescine level causes a partial blockade of the receptor. Hypoactivity in both sexes and spatial learning impairment in female *SSAT*-transgenic mice may also be associated with their altered hormone metabolism, leading to elevated circulating levels of adrenocorticotropin and corticosterone, and decreased levels of testosterone, thyroid-stimulating hormone and thyroxine [20]. Such hormonal changes can be attributable to the activation of the hypothalamic–pituitary axis that has been indicated in depression, behavioural abnormalities and learning disabilities.

Kidney and liver injury

Although putrescine is believed to protect the brain against injury, its function in other tissues is not similar. For instance, activation of SSAT and consequent putrescine accumulation is associated with liver and kidney ischaemia/reperfusion injuries in wild-type animals. When *SSAT*-knockout mice were subjected to occlusion and reperfusion protocols, they were significantly protected against injuries [21]. The results clearly suggest that lack of SSAT-mediated polyamine back-conversion is beneficial to these tissues. Whether the protection in *SSAT*-knockout animals is rather associated with lack of polyamine catabolism as such or with the maintenance of the levels of the higher polyamines is not clear. Renal spermidine and spermine levels were similar in both wild-type and knockout mice in response to the ischaemia/reperfusion procedure, whereas their hepatic levels were decreased substantially in the wild-type mice and slightly in *SSAT*-knockout mice [21]. These results indicate that the mechanism of tissue damage is different in the two tissues, albeit it appears to be mediated by SSAT in both. Eventually, the by-products of polyamine catabolism produced in PAO-catalysed reactions may play a significant role in tissue injuries.

We observed that partial hepatectomy activated polyamine catabolism and caused a failure to initiate liver regeneration in *SSAT*-transgenic rats when compared with non-transgenic littermates [22]. The transgenic rats showed accumulation of hepatic putrescine with a simultaneous profound decrease in spermidine and spermine contents. Apparently, putrescine alone was not sufficient to support liver regeneration. Pretreatment of the transgenic rats with α-methylspermidine prior to partial hepatectomy compensated for the loss of natural spermidine and restored liver regeneration, indicating that the regenerative process is indeed dependent on the presence of the higher polyamines [22]. Thus the methylated polyamine analogues may have potential in the treatment of liver traumas involving activated polyamine catabolism.

Keratinocyte dysfunction

The overaccumulation of putrescine is a critical factor in tumorigenesis, as already discussed above. Loss of hair after the first hair cycle has been shown in mice overexpressing *ODC* driven by the keratin promoter [23], or by *SSAT* driven by its own promoter [4] or by a metallothionein promoter [24]. The common denominator in these mice is the excessive accumulation of putrescine in their skin. Lan et al. [25] used a model in which *ODC* was driven by an involucrin promoter and, moreover, could be induced by virtue of oestrogen receptor ligand-binding protein fused to ODC. Induction of the fusion transgene with 4-hydroxytamoxifen resulted in putrescine accumulation and stimulated keratinocyte activation. Vascularization with increased expression of differentiation markers similar to those seen in wound healing was similarly observed in the transgenic skin [25].

The hairless phenotype of *SSAT*-transgenic mice is associated with early degeneration of hair follicles evident at the first follicle regression phase, catagen [26]. Dermal cysts and epidermal utriculi thereafter replace the normal hair follicles. Doubly transgenic mice, overexpressing both *ODC* and *SSAT*, exhibited an even higher accumulation of putrescine and more severe skin abnormalities than either singly transgenic line alone, again emphasizing the role of putrescine in this phenotype. In addition, inhibition of putrescine biosynthesis led to alleviation of the cutaneous changes and regrowth of hair [27]. Analysis of epidermal differentiation markers revealed that keratinocyte differentiation was clearly impaired in the transgenic mice, as well as in organotypic cultures of keratinocytes overexpressing *SSAT* [27]. These results demonstrate indisputably that polyamines, putrescine in particular, regulate the differentiation of keratinocytes.

Pancreatitis

Dramatic consequences following rapid depletion of spermidine and spermine are evident in transgenic animals in which the *SSAT* transgene is inducible by zinc through a metallothionein promoter and mainly expressed in pancreas and liver. Such rats develop acute necrotizing pancreatitis within 24 h after induction of transgene expression [28]. Contrary to conditions in brain and skin, where putrescine plays a central role, development of pancreatitis is primarily caused by depletion of spermidine and spermine in the transgenic animals. The pancreas is the richest source of spermidine in the body, and development of dramatic polyamine deficiency clearly precedes the onset of tissue damage. The strongest evidence in support of the role of polyamines in maintenance of pancreatic integrity comes from findings that metabolically relatively stable methylated polyamine analogues that fulfil the physiological roles of the natural polyamines prevent tissue damage caused by trypsinogen activation, an early step in acute pancreatitis, and protect animals from mortality that is associated with severe pancreatitis [22,29]. The potential contribution of hydrogen peroxide and reactive aldehyde products of the PAO-catalysed reaction to tissue damage in the transgenic model could be ruled out, as inhibition of PAO did not alleviate pancreatitis at all [28]. This model shows all of the general hallmarks of severe acute pancreatitis and could therefore be considered as a relevant model to study the pathogenesis of this disease. Activated polyamine catabolism leading to polyamine depletion has similarly been observed in standard experimental models of pancreatitis, induced by caerulein or L-arginine [29], or by taurodeoxycholate [30]. Activation of polyamine catabolism appears to be a common phenomenon in the pathogenesis of pancreatitis induced by different stimuli, although it may not be an equally important factor in each case. The final outcome is a result of all of the pathways activated after a given stimulus.

Diseases associated with an accelerated polyamine metabolic cycle

Links to Type 2 diabetes

A thorough investigation of the mechanisms leading to the fatless phenotype in *SSAT*-transgenic mice has introduced a novel concept of polyamine catabolism-related phenotypic changes in white adipose tissue. Fat tissue of these mice exhibited an highly increased putrescine pool, but the levels of other polyamines were only moderately depleted [31,32]. The increased activities of SSAT and the biosynthetic enzymes could be taken as evidence for an accelerated polyamine metabolic flux in these animals [32]. Indeed, this could be deduced from the depleted pools of ATP [31] and acetyl-CoA [32], both being immediately used in two reactions per each metabolic cycle depicted in Figure 2. The driving forces of this cycle thus are: (i) elevated SSAT activity, forced by transgene expression; (ii) activated ODC contributing to putrescine accumulation; and (iii) AdoMetDC activation facilitated by abundant putrescine and leading to adequate supply of dcAdoMet (decarboxylated AdoMet) for spermidine and spermine synthesis. The direct metabolic consequences of accelerated flux are: (i) excessive consumption of acetyl-CoA in SSAT-catalysed reactions; (ii) excessive consumption of ATP in the formation of AdoMet; and (iii) increased production of hydrogen peroxide and reactive aldehydes in PAO or SMO-catalysed reactions (Figure 2).

The downstream consequences of accelerated flux in a whole animal, such as seen in *SSAT*-transgenic mice, are far-reaching. Tissues exhibiting accelerated flux may suffer from energy shortage as such. Furthermore, rates of other metabolic pathways may be affected. In white adipose tissue, the elevated

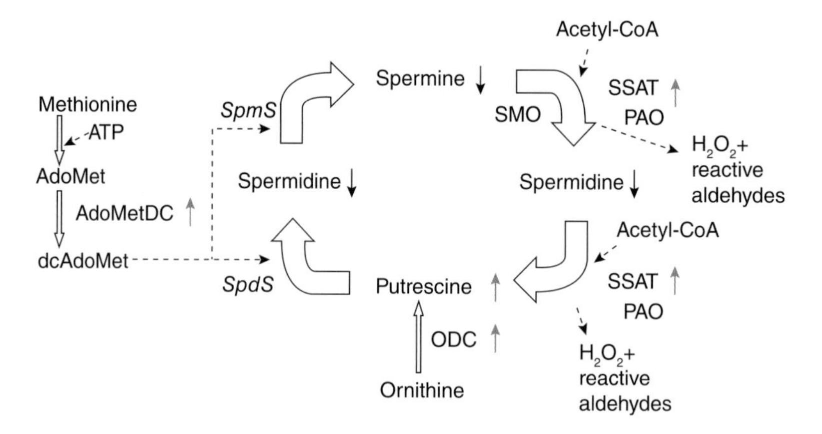

Figure 2. Accelerated metabolic cycle in *SSAT*-transgenic mice
The driving forces of the cycle are described in the text. Open arrows, enzymatic reactions; solid blue arrows up, overexpression, induction or accumulation; solid black arrows down, pool depletion; broken arrows, contribution of a molecule to a reaction.

AMP/ATP ratio appears to induce AMPK (AMP-activated protein kinase) which in turn activates PGC-1α [PPARγ (peroxisome-proliferator-activated receptor γ) co-activator 1α] and reduces formation of malonyl-CoA from acetyl-CoA by inhibiting acetyl-CoA carboxylase. Excessive consumption of acetyl-CoA in polyamine flux as such leads to its limited availability and thus partly also results in depletion of malonyl-CoA. As malonyl-CoA is a substrate of fatty acid biosynthesis and negatively regulates fatty acid oxidation, its depletion is manifested as decreased fatty acid synthesis and increased fatty acid oxidation [32]. PGC-1α is an important regulator of energy metabolism: it stimulates oxidative phosphorylation, thermogenesis, mitochondrial biogenesis, uncoupling, fatty acid oxidation and glucose transport. Most of these parameters were found to be increased in the fat tissue of *SSAT*-transgenic mice [31], as outlined in Figure 3. The fact that exogenously administered putrescine did not induce a similar expression profile in wild-type mice or glucose uptake in 3T3-L1 adipocytes ruled out the mechanisms that an elevated putrescine pool alone caused the striking phenotype of *SSAT* mice. As the metabolic profile of patients with Type 2 diabetes is totally opposite to that in *SSAT*-transgenic mice, the latter can be considered as a reverse model of the

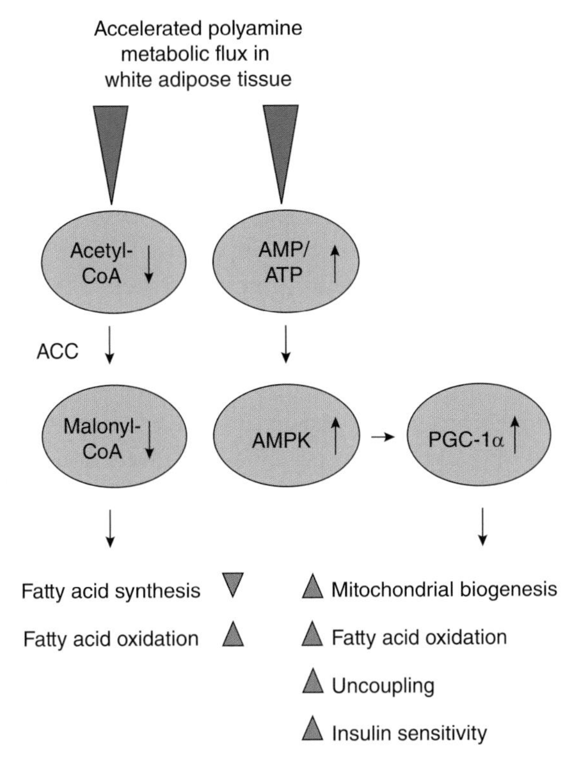

Figure 3. Metabolic consequences of an accelerated polyamine cycle in white adipose tissue

Consequences of accelerated flux are mediated by decreased acetyl-CoA and ATP pools. PGC-1α is the key mediator of the downstream affects of an elevated AMP/ATP ratio. ACC, acetyl-CoA carboxylase.

disease. Interestingly, and in support of our view, *SSAT*-deficient mice show an opposite metabolic profile when compared with *SSAT* mice. These mice gain fat and develop insulin resistance upon aging [32,33].

The rate of the metabolic cycle can be regulated by interfering with the enzymatic reactions involved in it. The flux in white adipose tissue was slowed down by inhibition of the *de novo* synthesis of putrescine by DFMO, which restored fat accumulation and the altered gene expression profiles [31]. Overexpressing *ODC* together with *SSAT* can also accelerate the cycle further. Doubly transgenic mice with *ODC* and *SSAT* overexpression governed by a metallothionein promoter were shown to exhibit enhanced flux in liver [34]. Since the same mice had more dramatic skin abnormalities [26], the effect of metabolic changes, such as depletion of ATP, and excessive production of toxic side-products cannot be excluded as factors contributing to the skin phenotype.

Tumorigenesis

Differential modulation of tumorigenesis by *SSAT* expression can be seen in two experimental approaches. In one approach, *SSAT*-transgenic mice were crossed with TRAMP (transgenic prostate adenocarcinoma model) mice. It appeared that tumour outgrowth and progression were suppressed by enhanced SSAT activity in doubly transgenic mice when compared with TRAMP mice [35]. The fact that the levels of the higher polyamine pools were minimally changed in the prostate of TRAMP/SSAT mice and the activities of the biosynthetic enzymes were increased, speaks for the mechanism that an enhanced polyamine cycle is the determinant of the outcome. This conclusion is also supported by the finding that, in the doubly transgenic mice, the prostatic levels of acetyl-CoA were significantly decreased in comparison with wild-type or TRAMP mice [35]. In another approach, *SSAT*-transgenic and knockout mice were crossed with intestinal tumorigenesis model $Apc^{Min/+}$ (MIN) mice. Unexpectedly, *SSAT*-overexpressing mice crossed with MIN mice developed more intestinal adenomas than MIN mice alone, whereas *SSAT*-knockout mice crossed with MIN mice developed fewer [36]. Again, despite overexpression of *SSAT* in the former hybrid, spermidine and spermine levels were not greatly affected in normal intestinal or colonic tissue or in tumours therein, but the biosynthetic enzymes were activated as an indication of an accelerated polyamine cycle. The opposite was observed in the hybrid lacking SSAT activity. In light of these findings, one may conclude that the differential effect of enhanced polyamine cycle on tumorigenesis in various tissues may depend on such determinants as accumulation of flux-related metabolites, sensitivity of cells to them, and the overall metabolic environment [36].

Reproductive impairment

Reproductive organs appear to be very sensitive to disturbed polyamine metabolism. Extremely high testicular overexpression of transgene-derived *ODC* and consequent overaccumulation of putrescine rendered a male founder mouse infertile, and also less pronounced overexpression in *ODC*-transgenic mice reduced reproductive performance of males due to disturbed spermatogenesis [37] as a result from altered spermatogonial DNA synthesis [38]. In comparison with *ODC*-transgenic males, the males overexpressing *SSAT* were completely fertile. This difference is most probably explained by a very moderate expression of the *SSAT* transgene and modest accumulation of putrescine in the testes of *SSAT*-transgenic males [4]. In contrast, the *SSAT*-overexpressing females developed ovarian hypofunction and uterine hypoplasia leading to infertility at a very young fertile age [4]. Interestingly, the expression levels of numerous genes were different in the uterus and ovary of transgenic females when compared with those in their non-transgenic littermates [39]. The exact roles of the differentially expressed genes in the observed phenotypic changes have not been clarified, but the findings tend to indicate that polyamines play an important role in controlling molecular mechanisms of reproductive tract development and function. In this context, it should again be borne to mind that the adult transgenic females had practically no visceral fat in their body. Therefore the possible effect of fatlessness in the hormonal regulation of reproduction in these mice cannot be ruled out.

Tissue-specific response to activated polyamine catabolism

Based on our findings with regard to the phenotypic changes in *SSAT*-transgenic mice (as described above), we can conclude that changes leading to pathogenesis in different tissues of these mice include distinctly disturbed homoeostasis, i.e. altered polyamine pattern, or virtually unaltered spermidine and spermine pools, but presumably enhanced flux or both. Activation of both biosynthetic enzymes, ODC and AdoMetDC, has been observed in all tissues studied, but the extent varies from tissue to tissue. This fact suggests the tissue-specific contribution as a driving force to the metabolic cycle. Among the tissues studied, some show clear-cut depletion of the higher polyamines (Table 2, categories A and B) to which the phenotypic changes may be primarily attributable. Other tissues with largely unaltered levels of the higher polyamines may be more prone to putrescine- and/or flux-mediated changes (Table 2, categories C and D).

Internal organs such as liver, pancreas, spleen, heart and kidney in adult *SSAT*-transgenic females and heart in young males were found to be enlarged [31]. Organomegaly is most likely a compensatory response to activated polyamine catabolism-caused stress. It is thus not surprising that activated polyamine catabolism in mice is associated with reduced life expectancy as compared with wild-type mice, and that the lifespan of doubly transgenic mice with *ODC* and *SSAT* overexpression is even more reduced [34].

Conclusions

The examples of the uses of transgenic and knockout models described in the present chapter show that these animals are relevant in the studies of polyamine

Table 2. Response of mouse tissues to ubiquitous overexpression of *SSAT* categorized according to the changes in polyamine pools

Polyamine levels in the tissues of *SSAT*-transgenic mice are expressed as changes in comparison with basal levels in non-transgenic mice.

Category	Change in polyamine pool	Mouse tissue
A	High putrescine accumulation Spermidine and spermine depletion	Spleen Small intestine Pancreas
B	Moderate putrescine accumulation Spermidine and spermine depletion	Liver Kidney Lung Thymus Heart
C	High putrescine accumulation Spermidine and spermine slightly affected	Skin Brain White adipose tissue
D	Moderate putrescine accumulation Spermidine and spermine slightly affected	Ovary Uterus Testis

functions, but also as models of pathological conditions and some specific human diseases. The exact mechanisms involved in these conditions still remain elusive in many cases. For generation of more sophisticated transgenic disease models, well-designed constructs with tissue-specific and preferably inducible or conditional expression/knockout are needed. Such models are currently being developed in our laboratory to target altered transgene expression more precisely. For instance, *SSAT* is being expressed exclusively in the pancreas by virtue of a conditionally activated elastase promoter or in adipose tissue by virtue of an adipose-tissue-specific fatty-acid-binding protein (aP2) gene promoter.

From a technical point of view, better methods of genetic engineering may replace the conventional pronuclear microinjection and knockout technology. A modern and much more efficient way of transgenesis is to use lentiviral vectors to carry transgenes into embryos via viral transduction. Gene silencing with the aid of short-interfering RNAs is a method of choice to study the loss-of-function, and it is also applicable to rats.

Summary

- *ODC or SSAT overexpression gives rise to a hairless phenotype that appears to be at least partially related to putrescine accumulation*

- *Induction of SSAT overexpression leads to delayed liver regeneration and development of acute pancreatitis following rapid depletion of the higher polyamines in SSAT-overexpressing rodents*
- *SSAT overexpression in fat tissue is associated with changes in glucose and energy metabolism that are regulated by accelerated polyamine metabolic flux, rather than by depleted polyamine pools*
- *Male reproduction is disturbed by ODC overexpression, whereas female reproduction is sensitive to SSAT overexpression*
- *SSAT expression modulates tumorigenesis differentially depending on the tissue*

References

1. Alhonen, L., Halmekytö, M., Kosma, V.-M., Wahlfors, J., Kauppinen, R. and Jänne, J. (1995) Life-long overexpression of ornithine decarboxylase (ODC) gene in transgenic mice does not lead to generally enhanced tumorigenesis or neuronal degeneration. Int. J. Cancer **63**, 402–404
2. Halmekytö, M., Alhonen, L., Alakuijala, L. and Jänne, J. (1993) Transgenic mice over-producing putrescine in their tissues do not convert the diamine into higher polyamines. Biochem. J. **291**, 505–508
3. Megosh, L., Gilmour, S.K., Rosson, D., Soler, A.P., Blessing, M., Sawicki, J.A. and O'Brien, T.G. (1995) Increased frequency of spontaneous skin tumors in transgenic mice which overexpress ornithine decarboxylase. Cancer Res. **55**, 4205–4209
4. Pietilä, M., Alhonen, L., Halmekytö, M., Kanter, P., Jänne, J. and Porter, C.W. (1997) Activation of polyamine catabolism profoundly alters tissue polyamine pools and affects hair growth and female fertility in transgenic mice overexpressing spermidine/spermine N^1-acetyltransferase. J. Biol. Chem. **272**, 18746–18751
5. Ikeguchi, Y., Wang, X., McCloskey, D.E., Coleman, C.S., Nelson, P., Hu, G., Shantz, L.M. and Pegg, A.E. (2004) Characterization of transgenic mice with widespread overexpression of spermine synthase. Biochem. J. **381**, 701–707
6. Korhonen, V.P., Niiranen, K., Halmekytö, M., Pietilä, M., Diegelman, P., Parkkinen, J.J., Eloranta, T., Porter, C.W., Alhonen, L. and Jänne, J. (2001) Spermine deficiency resulting from targeted disruption of the spermine synthase gene in embryonic stem cells leads to enhanced sensitivity to antiproliferative drugs. Mol. Pharmacol. **59**, 231–238
7. Lorenz, B., Francis, F., Gempel, K., Böddrich, A., Josten, M., Schmahl, W., Schmidt, J., Lehrach, H., Meitinger, T. and Strom, T.M. (1998) Spermine deficiency in Gy mice caused by deletion of the spermine synthase gene. Hum. Mol. Genet. **7**, 541–547
8. Feith, D.J., Shantz, L.M. and Pegg, A.E. (2001) Targeted antizyme expression in the skin of transgenic mice reduces tumor promoter induction of ornithine decarboxylase and decreases sensitivity to chemical carcinogenesis. Cancer Res. **61**, 6073–6081
9. Pegg, A.E., Feith, D.J., Fong, L.Y., Coleman, C.S., O'Brien, T.G. and Shantz, L.M. (2003) Transgenic mouse models for studies of the role of polyamines in normal, hypertrophic and neoplastic growth. Biochem. Soc. Trans. **31**, 356–360
10. Jänne, J., Alhonen, L., Keinänen, T.A., Pietilä, M., Uimari, A., Pirinen, E., Hyvönen, M.T. and Järvinen, A. (2005) Animal disease models generated by genetic engineering of polyamine metabolism. J. Cell. Mol. Med. **9**, 865–882
11. Jänne, J., Alhonen, L., Pietilä, M., Keinänen, T.A., Uimari, A., Hyvönen, M.T., Pirinen, E. and Järvinen, A. (2006) Genetic manipulation of polyamine catabolism in rodents. J. Biochem. (Tokyo) **139**, 155–160
12. Smith, M.K., Trempus, C.S. and Gilmour, S.K. (1998) Co-operation between follicular ornithine decarboxylase and v-Ha-ras induces spontaneous papillomas and malignant conversion in transgenic skin. Carcinogenesis **19**, 1409–1415

13. Lan, L., Trempus, C. and Gilmour, S.K. (2000) Inhibition of ornithine decarboxylase (ODC) decreases tumor vascularization and reverses spontaneous tumors in ODC/Ras transgenic mice. Cancer Res. **60**, 5696–5703
14. Feith, D.J., Origanti, S., Shoop, P.L., Sass-Kuhn, S. and Shantz, L.M. (2006) Tumor suppressor activity of ODC antizyme in MEK-driven skin tumorigenesis. Carcinogenesis **27**, 1090–1098
15. Nilsson, J.A., Keller, U.B., Baudino, T.A., Yang, C., Norton, S., Old, J.A., Nilsson, L.M., Neale, G., Kramer, D.L., Porter, C.W. and Cleveland, J.L. (2005) Targeting ornithine decarboxylase in Myc-induced lymphomagenesis prevents tumor formation. Cancer Cell **7**, 433–444
16. Wang, X., Levic, S., Gratton, M.A., Doyle, K.J., Yamoah, E.N. and Pegg, A.E. (2009) Spermine synthase deficiency leads to deafness and a profound sensitivity to α-difluoromethylornithine. J. Biol. Chem. **284,** 930–937
17. Halonen, T., Sivenius, J., Miettinen, R., Halmekytö, M., Kauppinen, R., Sinervirta, R., Alakuijala, L., Alhonen, L., MacDonald, E., Jänne, J. and Riekkinen, P. (1993) Elevated seizure threshold and impaired spatial learning in transgenic mice with putrescine overproduction in the brain. Eur. J. Neurosci. **5**, 1233–1239
18. Kaasinen, S.K., Gröhn, O.H., Keinänen, T.A., Alhonen, L. and Jänne, J. (2003) Overexpression of spermidine/spermine *N*[1]-acetyltransferase elevates the threshold to pentylenetetrazol-induced seizure activity in transgenic mice. Exp. Neurol. **183**, 645–652
19. Lukkarinen, J.A., Kauppinen, R.A., Gröhn, O.H.J., Oja, J.M.E., Sinervirta, R., Alhonen, L.I. and Jänne, J. (1998) Neuroprotective role of ornithine decarboxylase activation in transient cerebral focal ischemia: A study using ornithine decarboxylase-overexpressing transgenic rats. Eur. J. Neurosci. **10**, 2046–2055
20. Kaasinen, S.K., Oksman, M., Alhonen, L., Tanila, H. and Jänne, J. (2004) Spermidine/spermine *N*[1]-acetyltransferase overexpression in mice induces hypoactivity and spatial learning impairment. Pharmacol. Biochem. Behav. **78**, 35–45
21. Zahedi, K., Lentsch, A.B., Okaya, T., Barone, S.L., Sakai, N., Witte, D.P., Arend, L.J., Alhonen, L., Jell, J., Jänne, J. et al. (2009) Spermidine/spermine-*N*[1]-acetyltransferase ablation potects against liver and kidney ischemia reperfusion injury in mice. Am. J. Physiol. Gastrointest. Liver Physiol. **296**, G899–G909
22. Räsänen, T.L., Alhonen, L., Sinervirta, R., Keinänen, T., Herzig, K.H., Suppola, S., Khomutov, A.R., Vepsäläinen, J. and Jänne, J. (2002) A polyamine analogue prevents acute pancreatitis and restores early liver regeneration in transgenic rats with activated polyamine catabolism. J. Biol. Chem. **277**, 39867–39872
23. Soler, A.P., Gilliard, G., Megosh, L.C. and O'Brien, T. (1996) Modulation of murine hair follicle function by alterations in ornithine decarboxylase activity. J. Invest. Dermatol. **106**, 1108–1113
24. Suppola, S., Pietilä, M., Parkkinen, J.J., Korhonen, V.P., Alhonen, L., Halmekytö, M., Porter, C.W. and Jänne, J. (1999) Overexpression of spermidine/spermine *N*[1]-acetyltransferase under the control of mouse metallothionein I promoter in transgenic mice: evidence for a striking post-transcriptional regulation of transgene expression by a polyamine analogue. Biochem. J. **338**, 311–316
25. Lan, L., Hayes, C.S., Laury-Kleintop, L. and Gilmour, S.K. (2005) Suprabasal induction of ornithine decarboxylase in adult mouse skin is sufficient to activate keratinocytes. J. Invest. Dermatol. **124**, 602–614
26. Pietilä, M., Parkkinen, J.J., Alhonen, L. and Jänne, J. (2001) Relation of skin polyamines to the hairless phenotype in transgenic mice overexpressing spermidine/spermine *N*[1]-acetyltransferase. J. Invest. Dermatol. **116**, 801–805
27. Pietilä, M., Pirinen, E., Keskitalo, S., Juutinen, S., Pasonen-Seppänen, S., Keinanen, T., Alhonen, L. and Jänne, J. (2005) Disturbed keratinocyte differentiation in transgenic mice and organotypic keratinocyte cultures as a result of spermidine/spermine *N*-acetyltransferase overexpression. J. Invest. Dermatol. **124**, 596–601
28. Alhonen, L., Parkkinen, J.J., Keinänen, T., Sinervirta, R., Herzig, K.H. and Jänne, J. (2000) Activation of polyamine catabolism in transgenic rats induces acute pancreatitis. Proc. Natl. Acad. Sci. U.S.A. **97**, 8290–8295

29. Hyvönen, M.T., Herzig, K.H., Sinervirta, R., Albrecht, E., Nordback, I., Sand, J., Keinänen, T.A., Vepsäläinen, J., Grigorenko, N., Khomutov, A.R. et al. (2006) Activated polyamine catabolism in acute pancreatitis: α-methylated polyamine analogues prevent trypsinogen activation and pancreatitis-associated mortality. Am. J. Pathol. **168**, 115–122
30. Jin, H.T., Lämsä, T., Hyvönen, M.T., Sand, J., Räty, S., Grigorenko, N., Khomutov, A.R., Herzig, K.H., Alhonen, L. and Nordback, I. (2008) A polyamine analog bismethylspermine ameliorates severe pancreatitis induced by intraductal infusion of taurodeoxycholate. Surgery **144**, 49–56
31. Pirinen, E., Kuulasmaa, T., Pietilä, M., Heikkinen, S., Tusa, M., Itkonen, P., Boman, S., Skommer, J., Virkamäki, A., Hohtola, E. et al. (2007) Enhanced polyamine catabolism alters homeostatic control of white adipose tissue mass, energy expenditure, and glucose metabolism. Mol. Cell. Biol. **27**, 4953–4967
32. Jell, J., Merali, S., Hensen, M.L., Mazurchuk, R., Spernyak, J.A., Diegelman, P., Kisiel, N.D., Barrero, C., Deeb, K.K., Alhonen, L. et al. (2007) Genetically altered expression of spermidine/spermine N^1-acetyltransferase affects fat metabolism in mice via acetyl-CoA. J. Biol. Chem. **282**, 8404–8413
33. Niiranen, K., Keinänen, T.A., Pirinen, E., Heikkinen, S., Tusa, M., Fatrai, S., Suppola, S., Pietilä, M., Uimari, A., Laakso, M. et al. (2006) Mice with targeted disruption of spermidine/spermine N^1-acetyltransferase gene maintain nearly normal tissue polyamine homeostasis but show signs of insulin resistance upon aging. J. Cell. Mol. Med. **10**, 933–945
34. Suppola, S., Heikkinen, S., Parkkinen, J.J., Uusi-Oukari, M., Korhonen, V.P., Keinänen, T., Alhonen, L. and Jänne, J. (2001) Concurrent overexpression of ornithine decarboxylase and spermidine/spermine N^1-acetyltransferase further accelerates the catabolism of hepatic polyamines in transgenic mice. Biochem. J. **358**, 343–348
35. Kee, K., Foster, B.A., Merali, S., Kramer, D.L., Hensen, M.L., Diegelman, P., Kisiel, N., Vujcic, S., Mazurchuk, R.V. and Porter, C.W. (2004) Activated polyamine catabolism depletes acetyl-CoA pools and suppresses prostate tumor growth in TRAMP mice. J. Biol. Chem. **279**, 40076–40083
36. Tucker, J.M., Murphy, J.T., Kisiel, N., Diegelman, P., Barbour, K.W., Davis, C., Medda, M., Alhonen, L., Jänne, J., Kramer, D.L. et al. (2005) Potent modulation of intestinal tumorigenesis in $Apc^{min/+}$ mice by the polyamine catabolic enzyme spermidine/spermine N^1-acetyltransferase. Cancer Res. **65**, 5390–5398
37. Halmekytö, M., Hyttinen, J.-M., Sinervirta, R., Utriainen, M., Myöhänen, S., Voipio, H.-M., Wahlfors, J., Syrjänen, S., Syrjänen, K., Alhonen, L. and Jänne, J. (1991) Transgenic mice aberrantly expressing human ornithine decarboxylase gene. J. Biol. Chem. **266**, 19746–19751
38. Hakovirta, H., Keiski, A., Toppari, J., Halmekytö, M., Alhonen, L., Jänne, J. and Parvinen, M. (1993) Polyamines and regulation of spermatogenesis: selective stimulation of late spermatogonia in transgenic mice overexpressing the human ornithine decarboxylase gene. Mol. Endocrinol. **7**, 1430–1436
39. Min, S.H., Simmen, R.C., Alhonen, L., Halmekytö, M., Porter, C.W., Jänne, J. and Simmen, F.A. (2002) Altered levels of growth-related and novel gene transcripts in reproductive and other tissues of female mice over-expressing spermidine/spermine N^1-acetyltransferase (SSAT). J. Biol. Chem. **277**, 3647–3657
40. Halmekytö, M., Syrjänen, K., Jänne, J. and Alhonen, L. (1992) Enhanced papilloma formation in response to skin tumor promotion in transgenic mice overexpressing the human ornithine decarboxylase gene. Biochem. Biophys. Res. Commun. **187**, 493–497
41. Lukkarinen, J., Kauppinen, R.A., Koistinaho, J., Halmekytö, M., Alhonen, L. and Jänne, J. (1995) Cerebral energy metabolism and immediate early gene induction following severe incomplete ischaemia in transgenic mice overexpressing the human ornithine decarboxylase gene: evidence that putrescine is not neurotoxic *in vivo*. Eur. J. Neurosci. **7**, 1840–1849
42. Alhonen, L., Heikkinen, S., Sinervirta, R., Halmekytö, M., Alakuijala, P. and Jänne, J. (1996) Transgenic mice expressing the human ornithine decarboxylase gene under the control of mouse metallothionein I promoter. Biochem. J. **314**, 405–408
43. Soler, A.P., Gilliard, G., Megosh, L., George, K. and O'Brien, T.G. (1998) Polyamines regulate expression of the neoplastic phenotype in mouse skin. Cancer Res. **58**, 1654–1659

44. Guo, Y.J., Zhao, J.Q., Sawicki, J., Soler, A.P. and O'Brien, T.G. (1999) Conversion of C57Bl/6 mice from a tumor promotion-resistant to a -sensitive phenotype by enhanced ornithine decarboxylase expression. Mol. Carcinogen **26**, 32–36
45. Kilpeläinen, P.T., Saarimies, J., Kontusaari, S.I., Järvinen, M.J., Soler, A.P., Kallioinen, M.J. and Hietala, O.A. (2001) Abnormal ornithine decarboxylase activity in transgenic mice increases tumor formation and infertility. Int. J. Biochem. Cell Biol. **33**, 507–520
46. Shantz, L.M., Feith, D.J. and Pegg, A.E. (2001) Targeted overexpression of ornithine decarboxylase enhances β-adrenergic agonist-induced cardiac hypertrophy. Biochem. J. **358**, 25–32
47. Heljasvaara, R., Veress, I., Halmekytö, M., Alhonen, L., Jänne, J., Laajala, P. and Pajunen, A. (1997) Transgenic mice over-expressing ornithine and S-adenosylmethionine decarboxylases maintain a physiological polyamine homeostasis in their tissues. Biochem. J. **323**, 457–462
48. Nisenberg, O., Pegg, A.E., Welsh, P.A., Keefer, K. and Shantz, L.M. (2006) Overproduction of cardiac S-adenosylmethionine decarboxylase in transgenic mice. Biochem. J. **393**, 295–302
49. Fong, L.Y., Feith, D.J. and Pegg, A.E. (2003) Antizyme overexpression in transgenic mice reduces cell proliferation, increases apoptosis, and reduces N-nitrosomethylbenzylamine-induced forestomach carcinogenesis. Cancer Res. **63**, 3945–3954
50. Mackintosh, C.A., Feith, D.J., Shantz, L.M. and Pegg, A.E. (2000) Overexpression of antizyme in the hearts of transgenic mice prevents the isoprenaline-induced increase in cardiac ornithine decarboxylase activity and polyamines, but does not prevent cardiac hypertrophy. Biochem. J. **350**, 645–653
51. Kauppinen, L., Myöhänen, S., Halmekytö, M., Alhonen, L. and Jänne, J. (1993) Transgenic mice over-expressing the human spermidine synthase gene. Biochem. J. **293**, 513–516
52. Alhonen, L., Räsänen, T.L., Sinervirta, R., Parkkinen, J.J., Korhonen, V.P., Pietilä, M. and Jänne, J. (2002) Polyamines are required for the initiation of rat liver regeneration. Biochem. J. **362**, 149–153
53. Herzig, K.H., Jänne, J. and Alhonen, L. (2005) Acute pancreatitis induced by activation of the polyamine catabolism in gene-modified mice and rats overexpressing spermidine/spermine N^1-acetyltransferase. Scand. J. Gastroenterol. **40**, 120–121
54. Coleman, C.S., Pegg, A.E., Megosh, L.C., Guo, Y., Sawicki, J.A. and O'Brien, T.G. (2002) Targeted expression of spermidine/spermine N^1-acetyltransferase increases susceptibility to chemically induced skin carcinogenesis. Carcinogenesis **23**, 359–364
55. Guo, Y., Cleveland, J.L. and O'Brien, T.G. (2005) Haploinsufficiency for odc modifies mouse skin tumor susceptibility. Cancer Res. **65**, 1146–1149
56. Pendeville, H., Carpino, N., Marine, J.-C., Takahashi, Y., Muller, M., Martial, J.A. and Cleveland, J.L. (2001) The ornithine decarboxylase gene is essential for cell survival during early murine development. Mol. Cell. Biol. **21**, 6549–6558
57. Nishimura, K., Nakatsu, F., Kashiwagi, K., Ohno, H., Saito, T. and Igarashi, K. (2002) Essential role of S-adenosylmethionine decarboxylase in mouse embryonic development. Genes Cells **7**, 41–47

Index

A

C

D

E

F

G

H

I

K

L

M

N

O

P

R

S

T

U

W